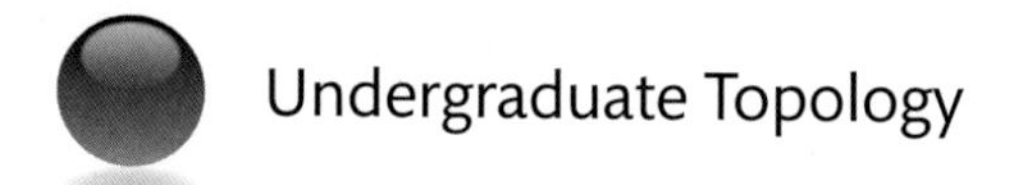

Undergraduate Topology

Undergraduate Topology

A Working Textbook

AISLING McCLUSKEY
Senior Lecturer in Mathematics
National University of Ireland, Galway

BRIAN McMASTER
Honorary Senior Lecturer
Queen's University Belfast

Great Clarendon Street, Oxford, OX2 6DP,
United Kingdom

Oxford University Press is a department of the University of Oxford.
It furthers the University's objective of excellence in research, scholarship,
and education by publishing worldwide. Oxford is a registered trade mark of
Oxford University Press in the UK and in certain other countries

First Edition published in 2014

Impression: 1

Published in the United States of America by Oxford University Press
198 Madison Avenue, New York, NY 10016, United States of America

British Library Cataloguing in Publication Data
Data available

Library of Congress Control Number: 2014930102

ISBN 978–0–19–870233–7 (hbk.)
ISBN 978–0–19–870234–4 (pbk.)

Printed and bound by
CPI Group (UK) Ltd, Croydon, CR0 4YY

Preface

This textbook offers an introduction at undergraduate level to an area known variously as general topology, point-set topology or analytic topology, *with a particular focus on helping students to build theory for themselves.* It emerged as a result of several years of our combined university teaching experience, stimulated by sustained interest in advanced mathematical thinking and learning, alongside established research careers in analytic topology. The modern-day student and instructor can benefit hugely from a contemporary text in the subject, a text that seeks to get to the heart of individual and independent mathematical learning. Our long experience suggests that there is a need for such a text.

Point-set topology is a discipline which needs relatively little background knowledge, but a good measure of determination to grasp the definitions precisely and to think and argue with straight and careful logic. It is widely acknowledged that the ideal way to learn such a subject is to teach it to yourself, proactively, by guided reading of brief skeleton notes and by doing your own spadework to fill in the details and to flesh out the examples. In reality, many students simply do not have the combination of time and commitment to implement this plan fully. A gentle touch is increasingly needed to engage students with the material and to build their confidence in handling it. Students undertaking this subject must come to terms with higher levels of abstraction and more sophisticated proofs than they are likely to have encountered before, and for such challenging notions it is invaluable to have an expert guide on hand. This text offers the possibility of step-by-step learning with relevant help, where required, in close proximity within the text. Thus it addresses a broad range of needs as reflected in the typically diverse student body: it facilitates the 'captive' student who is obliged to take the subject as a prerequisite for some other purpose, the student who does have time and commitment (even for a modest segment of the teaching term), and the instructor who would like to encourage the so-called Moore approach (again, even for part of the time or part of the class), and it can equally well be used for more conventional instructor-led courses.

The writing style is characterised by the use of accessible and engaging language, critically reminiscent in sufficient measure (not too much, not too little) of a workbook as opposed to the dense text that is the norm in other titles. The intention is that it will invite the reader/learner to use it and that, in particular, it will serve as a strong basis for self-directed study. Choice of content is almost inevitable for such a book; rather, it is how the content is presented to, and potentially experienced by, the learner that sets this text apart from the rightly revered classics of the area. Here we offer an alternative style of presentation, informed by long experience in the business of teaching and learning mathematics and also by research within the field of mathematics education, particularly in the area of advanced mathematical thinking. Even the visual appearance of a formal argument upon the printed page can now be a significant barrier to engagement for many. We have sought to overcome this by favouring an approach that mirrors the slow, careful process of developing a mathematical argument rather than opting for a more stylish monograph presentation. More specifically, it is our aim to facilitate a route to learning that enables comprehension, and therefore confidence, competence and enjoyment of the subject, and that wrests memorisation from its potentially dominant and devastating position in contemporary undergraduate mathematics education. The text offers content that is limited to the fundamental concepts of the subject, presented in such a fashion that students may achieve depth in their understanding of and facility with that content. The drive is for quality rather than quantity.

In essence, this is a text written by two mathematicians who are passionate both about the subject area and about teaching students. It is a result of long familiarity with the mathematical learning process, particularly in this subject area and with the contemporary student.

Contents

Comments to the instructor: what types of study this book will support

There are many excellent textbooks on general topology in print; what need is there of another?

The idea for this text arose from a related question that we occasionally asked ourselves: there are many excellent textbooks on general topology in print; why do we not teach out of one of them, rather than painstakingly creating our own lecture notes time after time?

A major part of the answer – for the present authors, anyway – lies in the expectations and preparedness of contemporary mathematics undergraduates, who, although just as skilled and insightful as ever in maths as such, tend to be short of experience in reading and writing extended prose passages of cumulative argument and – perhaps through a bias in the assessment system at post-primary level – over-reliant upon memory and memory-driven algorithms rather than upon fresh analysis of a question and the subsequent construction and communication of a persuasive answer to it. This is a double pity: for the mathematics graduate is virtually certain to be faced with problems for which no pre-learned algorithm will suffice, and much of the enjoyment and excitement of our discipline come from the struggle with such problems.

The ideal solution may very well be implementation of some version of the Moore method, in which students are presented only with the definitions and the statements of key results and examples, and encouraged to discover and develop their own arguments for what is asserted. Indeed, for a special class of undergraduate, this is a perfect mechanism for enhancing key skills of comprehension, construction of logical evidence and coherent communication. Yet, realistically, many students do not have the time, the motivation and the freedom from other competing demands on their energies to follow this path entirely. A text that encourages each of them to try this method for a proportion of their study, but that still offers a complete account of the material for when their time and energy run out, can bring benefits proportionate to the investment of effort that each finds it practicable to devote.

Our other key consideration was the choice of format in which to set forth that complete account, and here we confess to a degree of sympathy with students' common complaint that 'mathematics books are difficult to read'. Published mathematics for a professional readership is usually couched in lengthy, complex paragraphs of good literary style – and rightly so, given the appropriate sophistication of the argument and the target market – and similar eloquence is a frequent and enjoyable feature of the well-written textbook. Yet it seems to us important not to create the impression that learners are expected to do likewise: on the contrary, the aim that most of us hold for most of our students, as regards how they communicate their findings, is that they will develop a facility in proceeding stepwise through initially uncharted territory, verifying each point before going on to the next: bullet-point thinking, if not actually bullet-point writing. This expository style – at least as old as Euclid – has served us well for seventy generations, and it is also close to the manner in which modest mathematical discoveries are often made, especially in axiomatic work. With this in view, we have generally set out our demonstrations one step at a time, rather than in free-flowing prose. Our experience is that this makes it easier for the learner to compare his or her own attempts with model solutions and to comprehend such models where time pressures have prevented the preparation of personal attempts, that it sets a more realistic standard for the type of presentation that a diligent student should be producing by about the second or third draft of an assignment, and that it encourages systematic thinking both in discovery and in exposition. In this way, we seek to reduce the apparent gap between a learner's first rough-work exploration of an exercise and a submittable version, incorporating into the text aspects of teaching more usually to be found in tutorial sessions. The downside – that it uses up a little more paper than a compacted, fluid discourse would have done – may be considered a reasonable price to pay for the potential benefits.

So we have prepared a text that offers, in each chapter, an expository section that you could opt to make the raw material for a Moore treatment, and an 'Expansion' section giving complete specimen proofs – mostly in a logical but semi-informal bullet-point style such as a good student should aspire to create – that you could equally well interleave with the exposition to develop a conventional, tutor-led course. Where you position your delivery across that spectrum is for you to decide, but the book is designed to support your choice wherever it falls at each stage of the course. There is also an extensive Essential Exercises section, with model solutions to around half of the questions, so that you are resourced with a range of genuinely unseen assignments should that be appropriate. The introductory Chapter 1 summarises the background usually expected in such a class: essentially, an elementary treatment of metric spaces and the basics of analysis and of set theory that lead into such a study.

Comments to the student reader: how to use this book

Teaching topology over many years to a wide variety of students has convinced us that the perfect way to learn this material is not to be taught it at all. Learning it for yourself, through a guided reading of the key ideas and results, followed up by creating your own arguments to support what is claimed to be true – either as an individual or in small study groups – will give you a mastery of the subject, a depth of understanding, an ownership of the information that is very difficult to acquire quickly in any other way.

In the real world, however, most students simply do not have the combination of time and patience that full-blown self-tuition requires. What we have set out in the present book is a text that will allow you, the reader, to do as much self-tuition as will fit in with all the other demands on your attention, but provide a complete account that you, your class and your instructor can treat as a more conventional teaching textbook as and when necessary. Nevertheless, we strongly advise you to take the self-instruction route for at least part of the time, for the reasons outlined above.

Each chapter consists of two sections – which can usefully be thought of as Exposition and Expansion. The first part provides a collection of definitions, illustrations and results drawn from the standard content of a General Topology course, typically at final or pre-final undergraduate level. The second (and normally larger) part sets out supporting evidence for all of the assertions made in the first, excepting those that are genuinely immediate from the definitions. Generally speaking, this evidence is presented not in a finalised monograph format, but in a point-by-point fashion, rather as we would hope a diligent and time-rich student would set it out at about third draft. **Here is how we believe you should use each chapter for best effect:**

(a) Read carefully the first section. In particular, get the definitions very clear in your mind, using the examples to sharpen your comprehension of them.

(b) Make time to construct proofs of as many of the assertions – especially the easier ones – as you can. The material in this book is actually very self-contained, and all the ideas necessary for each demonstration should be in what you have just read (including the earlier chapters, of course), as long as you have acquired the necessary modest background (see below) and are prepared to argue logically.

(c) Check your demonstrations against those provided in the second (the 'Expansion') section of the chapter, and study the other proofs it provides for the results that you did not work out for yourself.

(d) Do several of the Essential Exercises. Again, check your solutions against the specimen ones where these have been covered in the 'Solutions to selected exercises'. Read – as time permits – some of the other Essential Exercises and model solutions.

(e) Proceed to the next chapter and repeat.

In connection with **background,** what a student wishing to study topology at this level needs to have got to grips with already is

(i) a decent understanding of basic real analysis;

(ii) a familiarity with the language of really elementary set theory (preferably, but not necessarily, including an encounter with the axiom of choice and Zorn's lemma);

(iii) an introductory course in metric spaces; and

(iv) a clear understanding of what counts as a proof in mathematics.

We do, in fact, revise these matters in Chapter 1, but not at sufficient length to give the reader the confidence in their use that is needed in order to make good progress.

A word about diagrams: most mathematicians working in this area, at whatever level of experience, draw rough-work diagrams all the time – as props to their intuition, as communication icons between themselves, as explanatory devices for their students and for half a dozen other reasons. We are keen to encourage students to doodle in this fashion at every stage of working through the book, provided only that it is understood that a diagram is not a proof. Please keep in mind that anything drawn upon a piece of paper, a blackboard or a whiteboard is necessarily living inside a two-dimensional Euclidean metric space, and that most of the interesting questions in topology reside in spaces whose structures are much more complicated than that. All the same, simple and properly interpreted diagrams are immensely valuable in grasping ideas and their interconnections, and will often serve as steps towards discovering a proof for oneself – as we hope to illustrate in a score of locations.

1 Introduction

Preamble

When you first began to study analysis – whether real or complex – you found that its central concerns were two distinct but closely interlinked ideas: limits and continuity.

Limits were concerned with approximation; they said that, by keeping on going, we could get better and better approximations to some ideal or ultimate position – indeed, approximations that came within **any** required tolerance from the ideal. There were, however, many apparently different types of limit: limits of sequences, sums to infinity of series, limits of functions $f(x)$ as $x \to \infty$ or as $x \to a$ or as $x \to a^{+} \ldots$, and later perhaps you met limits of sequences of functions, and wrestled with the subtle distinction between pointwise convergence and uniform convergence in this case. All of these ideas were given definitions that were recognisably similar in broad outline (along the lines of 'for all positive ε there exists positive δ such that, as soon as you cross the δ-threshold, all the approximations are good enough to pass the ε-test') but they differed in detail, depending on which variety of limit was being used. Most reasonable people are initially confused and irritated by these differences. (Is δ an integer or must it be allowed to be a real number this time? Is it $|x - a| < \delta$, or $a < x < a + \delta$, or $x > \delta$? Does it **really** matter whether one writes $\forall\ \varepsilon > 0\ \exists\ \delta > 0 \ldots$ or $\exists\ \delta > 0$ such that $\forall\ \varepsilon > 0 \ldots$?) Yet with practice, experience and the passing of time, most of these difficulties fade. Nevertheless, the striking resemblances between the results for these allegedly different limit concepts

(for instance,
$\lim(a_n + b_n) = \lim a_n + \lim b_n$,
$\sum(a_n + b_n) = \sum a_n + \sum b_n$,
$\lim_{x\to\infty}(f(x) + g(x)) = \lim_{x\to\infty} f(x) + \lim_{x\to\infty} g(x)$,
$\lim_{x\to a^-}(f(x) + g(x)) = \lim_{x\to a^-} f(x) + \lim_{x\to a^-} g(x)$)

leave behind a feeling that they are not really different at all, but rather that they are special cases of a more general definition which embraces all of them; and that if only we had the vision (and determination) to grasp this wider limit idea, all the particular results would fall into place as corollaries of theorems which we would then be able to prove about the general one.

Continuity, initially, was a more geometrical idea: it said that you could draw the graph of a function without lifting your pen off the page. Admittedly, that definition began to overstretch when you tried to apply it to functions such as

$$f(x) = x \sin(x^{-1}) \text{ if } x \neq 0,$$
$$f(0) = 0,$$

but it was a reasonably useful intuitive prop for functions having $\mathbb{R}$ or an interval in $\mathbb{R}$ as domain. Once you wanted or needed to examine functions that were continuous at some points but not at others, such as $[x]$, the integer part of x, or functions whose domain was not an interval, then a more precise and formal definition was required: and it was provided by the limit concept – a map f was to be considered as continuous on a set A if, for each point a_0 of A, the limit of $f(x)$ as $x \to a_0$ was $f(a_0)$, understanding that x has to belong to A here in order that $f(x)$ shall make sense. This allowed you to deal properly with functions such as the classic *Dirichlet function*

$$f(x) = \begin{cases} 1 & \text{if } x \text{ is rational,} \\ 0 & \text{if } x \text{ is irrational,} \end{cases}$$

which could then be seen as continuous on the rationals, and also continuous on the irrationals, and yet massively discontinuous on the reals. It also, of course, allowed you to prove results about continuity almost immediately using results on limits.

The above two paragraphs have been written on the assumption that the numbers involved were real; but virtually all the general remarks apply equally well to the complex case. (The only significant differences arise from the fact that the complex field is not ordered, and so one must avoid writing things like $\lim_{z \to z_0^+} f(z)$ or $z > z_0$ once z and z_0 are allowed to go complex.) Indeed, this discussion does not only apply to numbers: for instance, an algebraic or trigonometric expression in three variables can be viewed as a function whose domain is some subset of three-dimensional space, and it is both natural and straightforward to extend the ideas of limit and continuity so as to apply also in this arena. These remarks add force and breadth to the point we were making in the second paragraph of this chapter: there is much to be gained, in terms of efficiency and universality and understanding, if we can formulate a definition of limit which encompasses all the special cases that turn up in elementary analysis, and which can apply equally well to real numbers, complex numbers, points in space or any other *similar* class of objects. The subsidiary question raised by the last sentence is: what does *similar* mean in this context? What exactly is it about $\mathbb{R}, \mathbb{C}$ and $\mathbb{R}^3$ that permits such a discussion to take place?

Now, most of what we have said so far is an introduction, not to *something rich and strange*, but to something that you have already studied. *What it is* about $\mathbb{R}, \mathbb{C}$ and $\mathbb{R}^3$ is simply the fact that they are metric spaces. You will know from your study of these structures that as soon as we can spell out, in some reasonable fashion, what we mean by distances between objects in a set, then limits of sequences within that set make sense, and so does continuity of a function from that set to itself or to any other similarly structured set, and large swathes of elementary analysis transfer themselves painlessly to the new, wider context: that is, the same results are valid, and they are provable by the same methods as before. This is the origin – and, in part, the purpose – of metric space theory. (We shall briefly review it later in the present chapter.)

Yet there are circumstances in which limiting processes and continuous transformations make sense, but where the metric-space notion is inadequate to describe them. This is a central theme of topology, but one which it may be difficult to illustrate well on the basis of your previous study. Perhaps the following remarks will be sufficient for the moment. It is explicit in the formal definition of a metric space that (i) the distances from x to y and from y to x are equal; (ii) objects at zero distance are identical; and (iii) distances are always non-negative real numbers. However, any driver in a city with one-way streets knows that there are real-world situations which cannot be modelled using (i) as an axiom. Again, a useful measure of distance between continuous functions on a bounded closed interval $[a, b]$ is

$$d(f, g) = \int_a^b |f(x) - g(x)|\, dx;$$

yet some discontinuous functions can also be integrated, and one then finds that functions f and g that differ at only a **finite** number of values of x have $d(f, g) = 0$, although it would not then be true to say that $f = g$, despite the aspirations of (ii). Lastly, unbounded functions can often be approximated by using bounded ones, but a thoughtless and slavish use of metric concepts to express this approximation can give rise to 'infinite distance' assignments, contrary to (iii). Our overall point is that the idea of a metric space is just not subtle or general enough to provide a full and reliable description of every context in which limits and continuity are applicable. This is the task that general topology takes on. We shall exhibit a range of instances where it successfully accomplishes its mission.

Sets and mappings

We assume familiarity with very basic material on sets, such as:

1.1 Proposition (De Morgan's laws) For any family $\{B_i : i \in I\}$ of subsets of a given set X,

(i) $X \setminus \bigcup_{i \in I} B_i = \bigcap_{i \in I} (X \setminus B_i)$,

(ii) $X \setminus \bigcap_{i \in I} B_i = \bigcup_{i \in I} (X \setminus B_i)$, and

1.2 Proposition (the distributive laws) For any subset A of a given set X and any family $\{B_i : i \in I\}$ of subsets of X,

(i) $A \cap \bigcup_{i \in I} B_i = \bigcup_{i \in I} (A \cap B_i)$,

(ii) $A \cup \bigcap_{i \in I} B_i = \bigcap_{i \in I} (A \cup B_i)$.

Also assumed are the elementary ideas of mappings (which we feel free to call maps or functions without any implied change of emphasis) such as domain, codomain, range, one-to-one/1–1/injective, onto/surjective, bijective and inverses when they exist; note in particular, for a given map $f : X \to Y$, the notations

$$f(A) = \{f(a) : a \in A\},$$
$$f^{-1}(B) = \{x \in X : f(x) \in B\}$$

for each subset A of X and each subset B of Y. The notation $f^{-1}(B)$ is perhaps unfortunate, but in widespread use: it might tempt you to think that f has an inverse mapping, but this is not generally the case (indeed, it is expressly *not* the case unless f is both 1–1 and onto). The essential thing to remember about the *sets* $f(A) \subseteq Y$ and $f^{-1}(B) \subseteq X$ is which *points* are in them:

$$y \in f(A) \Leftrightarrow \exists x \in A \text{ such that } f(x) = y,$$
$$x \in f^{-1}(B) \Leftrightarrow f(x) \in B.$$

As processes on sets, f and f^{-1} behave rather nicely; we recall the following easy but vital rules.

1.3 Proposition

(a) Let $f: X \to Y$ and let $\{A_i : i \in I\}$ be a family of subsets of X. Then:
   - (i) $f(\bigcup\{A_i : i \in I\}) = \bigcup\{f(A_i) : i \in I\}$,
   - (ii) $f(\bigcap\{A_i : i \in I\}) \subseteq \bigcap\{f(A_i) : i \in I\}$,
   - (iii) equality does not usually obtain in (ii).

(b) Let $f: X \to Y$ and let $\{B_i : i \in I\}$ be a family of subsets of Y. Then:
   - (i) $f^{-1}(\bigcup\{B_i : i \in I\}) = \bigcup\{f^{-1}(B_i) : i \in I\}$,
   - (ii) $f^{-1}(\bigcap\{B_i : i \in I\}) = \bigcap\{f^{-1}(B_i) : i \in I\}$,

   and also
   - (iii) $f^{-1}(Y \setminus B) = X \setminus f^{-1}(B)$ for any $B \subseteq Y$.

(c) Let $f: X \to Y$ and $A \subseteq X, B \subseteq Y$. Then:
   - (i) $f(f^{-1}(B)) \subseteq B$,
   - (ii) $A \subseteq f^{-1}(f(A))$,
   - (iii) equality obtains in (i) for all B if and only if f is onto,
   - (iv) equality obtains in (ii) for all A if and only if f is 1–1.

On the infrequent occasions when a function has to be described without needing to acquire a referring symbol, we use the $x \mapsto y$ notation. For example, a passing reference to the general quadratic (real) function can be made by writing 'the map $x \mapsto ax^2 + bx + c$ from $\mathbb{R}$ to $\mathbb{R}$'.

We assume that you are confident about the distinction between finite sets and infinite sets. (Note in particular that the empty set counts as finite.) We call a set **countably infinite** if there is a bijection between it and the set $\mathbb{N}$ (less formally, if its elements can be listed as the terms of an endless list, an (infinite) sequence). By a **countable** set we intend one that is either finite or countably infinite. **Uncountable** means not countable. Keep in mind that $\mathbb{Q}$ is countable, that any interval in $\mathbb{R}$ is uncountable (with the obvious exception of degenerate one-element intervals of the form $[a, a]$), that the union of countably many countable sets is countable and that the (Cartesian) product of a finite number of countable sets is countable (we shall review the idea of product set in Chapter 5).

Be clear about the difference between element and subset, especially when dealing with a set of sets. For instance, if Γ is the collection of all open circular discs in the coordinate plane, and Δ is the set of open circular discs centred on the origin and of rational radius, and D is the particular disc $B(0, 1)$ whose centre is the origin and whose radius is 1, then D is an element of Δ but not a subset, Δ is a subset of Γ but not an element, and so on. These distinctions may seem petty (and *academic* in the derogatory sense!) but they really do matter. Be careful, therefore, always to use the symbols $\subseteq$ and $\in$ correctly.

Zorn's lemma

The one somewhat sophisticated idea from set theory that we very occasionally need to use is Zorn's lemma, and we shall set out the lengthy but easy details of its background here.

Recall that, if X is a set and $\leq$ is a relation on X, then:

- $\leq$ is a *partial order* if it is reflexive, anti-symmetric and transitive

  (that is, $x \leq x \ \forall \ x \in X$,

  $(x \leq y$ and $y \leq x) \Rightarrow x = y$,

  $(x \leq y$ and $y \leq z) \Rightarrow x \leq z)$;
- $\leq$ is a *total order* if, in addition to being a partial order, it satisfies the condition that every two elements of X are comparable; that is,

  given $x, y \in X$, either $x \leq y$ or $y \leq x$;
- $\leq$ is a *well-ordering* if, in addition to being a total order, it satisfies the condition:

  every non-empty subset of X has a least member

  (that is, $\emptyset \neq A \subseteq X$ implies $\exists \ a_0 \in A$

  such that, for every $a \in A, a_0 \leq a)$;
- whenever $x \leq y$ and $x \neq y$, it is customary to write $x < y$. The notations $x < y$ and $y > x$ are interchangeable, as are the notations $x \leq y$ and $y \geq x$.

By way of illustration, set inclusion $\subseteq$ is a partial order on any collection of sets because the three necessary rules

$$A \subseteq A \text{ always,}$$
$$(A \subseteq B \text{ and } B \subseteq A) \Rightarrow A = B,$$
$$(A \subseteq B \text{ and } B \subseteq C) \Rightarrow A \subseteq C$$

are obviously satisfied. Again, on the set $\mathbb{N}$ the relation *div* defined by

$$x \; div \; y \Leftrightarrow x \text{ is a factor of } y$$

is a partial order because

$$x \; div \; x \text{ always,}$$
$$(x \; div \; y \text{ and } y \; div \; x) \Rightarrow x = y,$$
$$(x \; div \; y \text{ and } y \; div \; z) \Rightarrow x \; div \; z$$

evidently hold. Neither of these is (generally) a total order: for example, neither 6 *div* 10 nor 10 *div* 6 is true, and neither $\{1,2\} \subseteq \{2,3\}$ nor $\{2,3\} \subseteq \{1,2\}$ is true. Furthermore, the *usual order* $\leq$ on $\mathbb{N}$ or $\mathbb{Q}$ or $\mathbb{R}$ is obviously a total order, and the usual order on $\mathbb{N}$ is even a well-ordering, but the usual order on $\mathbb{Q}$ is not a well-ordering (what is the least of the positive rationals whose squares exceed 2? there isn't one!) and the usual order on $\mathbb{R}$ is not a well-ordering (there is, for instance, no least member of the open interval $(0, 1)$). Note also that a set which has a partial order is called a *poset*, and that a set with a total order is called a *chain*. A set with a well-ordering is just called a *well-ordered set*.

Any non-empty subset A of a poset $(X, \leq)$ can be regarded as a poset in its own right if we merely restrict the ordering on X to apply only to the elements of A. If X is a poset but is not totally ordered, there will certainly be subsets of X which are totally ordered. These are called *chains in* X. For instance, $\{3, 6, 30, 120, 1200\}$ is a chain in $(\mathbb{N}, div)$ and $\{\{a\}, \{a, b\}, \{a, b, c, d\}, \{a, b, c, d, e, f\}\}$ is a chain in the subsets of the alphabet under set inclusion.

If A is a subset of a poset $(X, \leq)$ and y is an element of X, we call y an **upper bound** of A if, for every $a \in A$, $a \leq y$.

If A is a subset of a poset $(X, \leq)$ and y is an element of A (note: of A this time!), we say that y is a maximal element of A if $(a \in A, y \leq a)$ together imply that $a = y$: that is, if no element of A is **strictly greater** than y. This is not the same as greatest element: although the greatest element of A (if it exists) will certainly be a (unique) maximal element of A, sets can readily have many different maximal elements (none of which could therefore be a greatest element for that set). For example, when $\{3, 4, 5, 6, 7, 8, 14\}$ is ordered by *div*, 5 and 6 and 8 and 14 are maximal elements for it.

Now that the undergrowth has been cleared, we can at last state Zorn's lemma.

1.4 Zorn's lemma If, in a poset, *every* chain has an upper bound, then the poset has at least one maximal element.

The proof of this assertion lies outside the scope of the present text, but we shall demonstrate here a small example of how to use the lemma.

1.5 Exercise Show that there is, in the coordinate plane $\mathbb{R}^2$, a subset D with the following two properties:

(1) no three points of D are collinear, and

(2) every point of $\mathbb{R}^2 \setminus D$ lies on the line through two points of D.

Solution Let us call $B \subseteq \mathbb{R}^2$ *line-shy* if no three of its points are collinear. Let $\mathcal{L}$ be the collection of all line-shy subsets. Then $(\mathcal{L}, \subseteq)$ is a poset. We seek to apply Zorn's lemma to it.

Let $\mathcal{T}$ be any chain in $(\mathcal{L}, \subseteq)$, that is, $\mathcal{T}$ is a family $\{C_i : i \in I\}$ of line-shy sets every two of which are comparable (under $\subseteq$). What can we say about their union $K = \bigcup\{C_i : i \in I\}$? Any three elements p, q, r of K will need to belong to three sets $C_{i(p)}, C_{i(q)}, C_{i(r)}$ within $\mathcal{T}$, and the *comparability* clause says that one of those three sets – let us call it C_j – contains the other two. So p and q and r all lie in the line-shy set C_j, and cannot therefore be collinear. Thus, K is *in* $\mathcal{L}$, and is an upper bound in $(\mathcal{L}, \subseteq)$ for the entire chain $\mathcal{T}$.

Now that we have seen that Zorn's lemma can be applied to this poset, it assures us that $\mathcal{L}$ has a maximal element D. Since D *is* an element of $\mathcal{L}$, (1) holds. Since it is maximal, any attempt to augment D even by the adjunction of one extra element will push it outside $\mathcal{L}$. That is, for any $s \in \mathbb{R}^2 \setminus D$, $D \cup \{s\}$ does not belong to $\mathcal{L}$, and three of its points must therefore be collinear. The three points cannot all belong to D (since D is line-shy), so s is one of the three. This proves (2).

Actually, the above proof is a classic case of the use of a sledgehammer to crack a walnut, because a circular or elliptical contour would do the job of the set D as described in 1.5 without any appeal to Zorn! Nevertheless, our new sledgehammer will crack tougher nuts than this: for instance, almost exactly the same argument will produce a subset of the plane satisfying (1) and (2) *that contains any given line-shy subset*, or will let us carry out the whole exercise *within any given subset of the plane*. More importantly, this argument shows how, in practice, Zorn is usually applied, and we shall use it later in at least one scenario where no quick and easy solution is available.

Zorn's lemma is closely associated with another major and sophisticated result in set theory which it may be useful to mention at this stage:

1.6 The well-ordering theorem Let X be any set. Amongst all the relations on X, there is at least one that is a well-ordering: that is, *any set can be well-ordered.*

(The demonstration that 1.6 may be deduced from 1.4 also lies well outside the scope of the present text. We emphasise that, if the given set X happens to come with a (partial or total) order already in place, then the well-ordering guaranteed here may well be utterly unrelated to it.)

The least uncountable ordinal

One of the immediate consequences of 1.6 is that there exist uncountable well-ordered sets. For instance, we could just impose a well-ordering on $\mathbb{R}$, and it is then a slight additional technicality to arrange that, under this well-ordering, the set shall have a greatest element: let us call it m for the moment. Under this well-ordering (which, we should again emphasise, will have no relationship at all with the natural ordering of the real numbers), $\mathbb{R}$ will possess a least element, conventionally denoted by 0. We may use the familiar notation of intervals in this well-ordered set so that, for example, $[0, m]$ is the whole of $\mathbb{R}$, and we see that there exist values of $x \in \mathbb{R}$ for which the so-called *initial segment* $[0, x)$ is uncountable. The well-ordering now assures us that there is a *least* such x: so, denoting it by ω_1, we have that $[0, \omega_1)$ is uncountable but, for every $t < \omega_1$, the initial segment $[0, t)$ is countable.

The well-ordered set $[0, \omega_1)$ we have just been describing is of considerable importance in topology (and certainly elsewhere in mathematics) and it is essentially unique in the sense that, if W_1 and W_2 are two well-ordered sets that are uncountable and have all of their initial segments countable, then there is a bijection $f : W_1 \to W_2$ that preserves the ordering (that is to say, $x < y$ in W_1 if and only if $f(x) < f(y)$ in W_2). It is called, for obvious reasons, the *least uncountable ordinal.* From our point of view, the most important of its properties is the following:

1.7 Proposition Any countable subset of the least uncountable ordinal has an upper bound.

Let us be more specific as to why this result interests us in the present context. If, in the notation above, $C \subseteq [0, \omega_1)$ and C is countable, then by 1.7 there exists $t < \omega_1$ such that for every $c \in C$ we have $c \leq t$. That means that the interval (t, ω_1) contains none of the elements of C. In effect, if we now restore ω_1 as top element to our ordered set and choose to work in $[0, \omega_1]$ rather than in $[0, \omega_1)$, no sequence in $[0, \omega_1)$ can get close to ω_1, because some neighbourhood $(t, \omega_1]$ screens ω_1 off from all the terms in the sequence: for the range of the sequence is, of course, a countable set. We are, of course, getting too far ahead of our exposition now, for we have not yet defined neighbourhood of a point nor limit of a sequence in ways that are meaningful in such a context as this; but now we shall move towards remedying this defect through a brief revision of the elements of metric space theory that the reader of such a text as this is expected to have encountered already.

Metric spaces

A metric on a non-empty set is an assignment of a real distance to each pair of elements of the set that mirrors, in the following sense, how physical distances behave:

1.8 Definition A metric d on a non-empty set M is a mapping from $M \times M$ into the real line $\mathbb{R}$ such that, for all x, y, z in M:

(i) $d(x, y) \geq 0$,

(ii) $d(x, y) = d(y, x)$,

(iii) $d(x, y) = 0$ if and only if $x = y$ and

(iv) $d(x, z) \leq d(x, y) + d(y, z)$.

Then the pair of objects (M, d) is called a *metric space.*

1.9 Examples

(i) Any non-empty subset of the real line or the coordinate plane or ordinary three-dimensional space or, indeed, of 'ordinary' n-dimensional space becomes a metric space when we choose to measure distance in the familiar Euclidean way. Furthermore, there are non-Euclidean metrics even on such everyday spaces as these: for instance, both

$$d'((x_1, y_1), (x_2, y_2)) = |x_1 - x_2| + |y_1 - y_2|$$

and

$$d''((x_1, y_1), (x_2, y_2)) = \text{Max}\{|x_1 - x_2|, |y_1 - y_2|\}$$

also define useful and natural metrics d', d'' on the coordinate plane.

(ii) Any non-empty set at all may be made into a so-called *discrete* metric space by declaring the distance between every two distinct points to be 1.

(iii) The family of continuous real-valued functions on $[0, 1]$ becomes a metric space when we define the function-to-function distance as $d(f, g) = \int_0^1 |f(x) - g(x)|\, dx$, and it becomes a significantly different metric space when, instead, we define distance as $D(f, g) = \text{Max}\{|f(x) - g(x)| : 0 \leq x \leq 1\}$.

(iv) In contrast to what usually happens in algebraic areas, where it is only certain special subsets of a structure (subgroups, subfields, sub-vector spaces and so on) that can be made into a substructure under the same processes, absolutely any non-empty subset A of a metric space (M, d) can be made into a sub-metric space merely by restricting the metric d to apply only to pairs of points from A. When this is done, A is called simply a *subspace* of (M, d). This will generate a wide variety of metric spaces starting off from almost any standard example, as we have already indicated in (i) above. By way of illustrating how wide the divergence in behaviour of a subspace from its parent space can be, we now describe an example which will prove useful at several points later in the text.

1.10 Example: the Cantor excluded-middle sets Whenever A is a set of real numbers consisting of the union of finitely many pairwise disjoint bounded closed intervals

$$A = [a_1, b_1] \cup [a_2, b_2] \cup [a_3, b_3] \cup \cdots \cup [a_n, b_n],$$

let us define the set $^*A$ to be the union of the top and bottom one-tenths of each of these intervals:

$$^*A = \left[a_1, a_1 + \frac{1}{10}(b_1 - a_1)\right] \cup \left[a_1 + \frac{9}{10}(b_1 - a_1), b_1\right] \cup \cdots$$
$$\cdots \cup \left[a_n + \frac{9}{10}(b_n - a_n), b_n\right].$$

Since the set $^*A$ is itself the union of a similar family of intervals, we may apply $*$ to it in its turn to create $^{**}A$ and, further, $^{***}A$ and so on.

The case which most directly concerns us is that in which we begin this process with $[0, 1]$, because standard decimal notation then gives us a simple way to determine exactly which numbers belong to $^*[0, 1]$ and all the later iterates. In $^*[0, 1]$, we find exactly those numbers that can be given decimal expansions with only 0 or 9 in the first decimal place. This is because, in removing the central eight-tenths of the interval, we have taken out those whose first decimal digit was $1, 2, 3, \cdots$ or 8. Notice the importance of the word *can*: for instance, $\frac{1}{10}$ is usually expressed as the decimal 0.1, but it *can* be written as $0.0\dot{9}$, which agrees with its inclusion in $^*[0, 1]$. For the same reason, $^{**}[0, 1]$ comprises those real numbers that can be given a decimal expansion with only 0 or 9 in the first two places, $^{***}[0, 1]$ comprises those whose expansions can be done with only 0 or 9 in the first three places, and so on.

Having thus generated an infinite nested sequence

$$[0,1] \supseteq {}^*[0,1] \supseteq {}^{**}[0,1] \supseteq {}^{***}[0,1] \supseteq \cdots \supseteq {}^{*^n}[0,1] \supseteq \cdots$$

of sets on the real line, it is natural to ask: what is their intersection? (Let us denote it by $Cantor_{10}$.) Arithmetically, the answer is obvious from the discussion above: $Cantor_{10}$ comprises exactly those real numbers whose *entire* decimal expansions can be expressed using only 0s and 9s. The detailed structure of the metric space $Cantor_{10}$ is, however, less obvious and more interesting, as we shall see.

The bio-historical accident that humans usually do their arithmetic to base 10 has no essential role to play in the above construction: we could have opted to work to any other base b greater than 2 without significantly affecting the structure of the resulting space (let us denote it by $Cantor_b$), as we shall make clear in Chapter 3. Had we chosen, for example, to count to base 3, we should have removed central thirds from the various intervals instead of central eight-tenths, and 'tresimal' expansion of real numbers then gives us a convenient way to keep track of what has been excised and what ultimately remains: because now, $^*[0, 1]$

comprises the real numbers whose tresimal representation can be done without using 1 as the first digit, $^{**}[0,1]$ comprises those whose representations can avoid using 1 as the first or second digit, and so on, while *Cantor*$_3$ itself consists of all real numbers that are expressible in the form

$$0.a_1a_2a_3a_4\cdots{}_{[3]} = \sum \frac{a_n}{3^n},$$

where each base-3 digit a_n is either 0 or 2. It is this version of the Cantor set that is most commonly presented, and from which its customary name of *middle thirds set* is derived.

1.11 Exercise

(i) Show that the average of any two distinct elements of *Cantor*$_5$ lies outside *Cantor*$_5$.

(ii) Verify that the corresponding statement for *Cantor*$_3$ is false.

1.12 Definition Given a metric space (M, d):

(i) an *open ball* is a subset of the form $B(x, r) = \{y \in M : d(x, y) < r\}$ for some $x \in M$ and real $r > 0$;

(ii) an *open set* is a set that can be expressed as a union (finite or infinite) of open balls;

(iii) a *closed set* is a subset of M whose complement is an open set.

It is evident that M itself and the empty set are open and that the union of an arbitrary family of open sets is open, and it is an easy induction argument to check that the intersection of any finite family of open sets is open. Be aware that a set can be both open and closed (for instance, in a discrete space, every set is) and that many (most?) sets are neither open nor closed.

To bring out the connections with the historical roots of this material, we can opt to define continuity and sequential limits for metric spaces exactly as one does in elementary analysis, thus:

1.13 Definitions

(i) Given a pair of metric spaces (M, d) and (M', d') and a mapping $f : M \to M'$, we say that f is *continuous* if, for each $x \in M$ and each positive real number ε, there is a positive real number δ such that

$$y \in M, d(x, y) < \delta \text{ together imply } d'(f(x), f(y)) < \varepsilon.$$

(ii) Given a metric space (M, d), a sequence $(x_n)_{n\in\mathbb{N}}$ in M and an element x of M, we say that $(x_n)_{n\in\mathbb{N}}$ *converges* to the *limit* x (and we write $x_n \to x$) if, for each positive real number ε, there is a positive integer n_0 such that

$$n \in \mathbb{N}, n \geq n_0 \text{ together imply } d(x_n, x) < \varepsilon.$$

However, in practice it is often more convenient to redefine ideas such as these in terms not of distance but of open sets and/or open balls. (Such characterisations tend to be easier to handle, in part because the fundamental behaviour of open sets is simpler than that of distance measurements.) For example:

1.14 Lemmas

(i) Given a pair of metric spaces (M, d) and (M', d') and a mapping $f : M \to M'$, then f is continuous if and only if:

$$f^{-1}(G) \text{ is open in } (M, d) \text{ whenever } G \text{ is open in } (M', d').$$

(ii) Given a metric space (M, d), a sequence $(x_n)_{n \in \mathbb{N}}$ in M and an element x of M, then $(x_n)_{n \in \mathbb{N}}$ converges to x if and only if:

given open set G such that $x \in G$, we have $x_n \in G$ for all sufficiently large n.

(Incidentally, in (ii) above we may replace 'open set' by 'open ball' or by 'open ball centred on x' and the result remains valid.)

Continuing for a moment with the theme of equivalent definitions, we point out that sequential convergence in its turn will serve to provide alternative definitions for many of the ideas in metric spaces; witness the following few results:

1.15 Lemmas

(i) Given a pair of metric spaces (M, d) and (M', d') and a mapping $f : M \to M'$, then f is continuous if and only if:

$$\text{whenever a sequence } x_n \to x \text{ in } (M, d), \text{ then } f(x_n) \to f(x) \text{ in } (M', d').$$

(ii) Given a metric space (M, d) and a subset A of M, then A is a closed set if and only if:

whenever a sequence $(x_n)_{n \in \mathbb{N}}$ of elements of A converges to x in (M, d), then $x \in A$.

1.16 Lemma We consider the following statements about a non-empty subset A of a metric space (M, d):

(a) A is closed and bounded;

(b) within every family of open sets whose union contains A, there is a finite subfamily whose union already contains A;

(c) every sequence in A has a subsequence that converges to an element of A.

Then:

(i) in all cases, (b) and (c) are equivalent, and they imply (a);

(ii) in the Euclidean spaces, all three are equivalent.

Proofs of these results will be found in any book that deals with the elementary theory of metric spaces, so we shall not include them here. Presently our main purpose – apart from reminding the reader of material almost surely encountered earlier, and with a view to accessing easy illustrative examples – is to flag up the fact that once we move into topology proper and lose the metric, the various equivalences that are so much a feature of metric space theory and practice can no longer be relied upon. In some cases (1.15, for instance), what was in the metric setting a necessary and sufficient condition will turn out to be only necessary in topology; in others (1.16's (b) if and only if (c) illustrates this), it will not even be that; in others still (consider, for example, (b) implies (a) in 1.16), it may lose meaning altogether as soon as there is no longer a metric.

Expansion of Chapter 1

1.1

(i)

$$x \in \text{RHS} \Leftrightarrow (\forall i \in I)\ x \in X \setminus B_i$$
$$\Leftrightarrow (\forall i \in I)\ x \notin B_i$$
$$\Leftrightarrow x \notin \bigcup_{i \in I} B_i$$
$$\Leftrightarrow x \in X \setminus \bigcup_{i \in I} B_i = \text{LHS}.$$

(ii)

$$x \in \text{LHS} \Leftrightarrow x \notin \bigcap_{i \in I} B_i$$
$$\Leftrightarrow \exists i \in I \text{ such that } x \notin B_i$$
$$\Leftrightarrow \exists i \in I \text{ such that } x \in X \setminus B_i$$
$$\Leftrightarrow x \in \bigcup_{i \in I} (X \setminus B_i) = \text{RHS}.$$

1.2

(i)

$$x \in \text{LHS} \Leftrightarrow x \in A \text{ and } \exists i \in I \text{ such that } x \in B_i$$
$$\Leftrightarrow \exists i \in I \text{ such that } x \in A \cap B_i$$
$$\Leftrightarrow x \in \bigcup_{i \in I} (A \cap B_i) = \text{RHS}.$$

(ii) Similar.

1.3

(a) (i) (For $y \in Y$:)

$$y \in \text{LHS} \Leftrightarrow \exists x \in \bigcup_{i \in I} A_i \text{ such that } y = f(x)$$
$$\Leftrightarrow \exists i \in I, x \in A_i \text{ such that } y = f(x)$$
$$\Leftrightarrow \exists i \in I \text{ such that } y \in f(A_i)$$
$$\Leftrightarrow y \in \bigcup_{i \in I} f(A_i) = \text{RHS.}$$

(ii)

$$y \in \text{LHS} \Leftrightarrow \exists x \in \bigcap_{i \in I} A_i \text{ such that } y = f(x)$$
$$\Leftrightarrow \exists x \text{ such that}, \forall i \in I, x \in A_i \text{ and } y = f(x)$$
$$\Rightarrow (\forall i \in I)\ y \in f(A_i)$$
$$\Leftrightarrow y \in \bigcap_{i \in I} f(A_i) = \text{RHS.}$$

(iii) For instance, define $f : \mathbb{R} \to \mathbb{R}$ by $f(x) = x^2$, take $A_1 = [-2, 0], A_2 = [0, 3]$.

$$\text{Then } f(A_1) \cap f(A_2) = [0,4] \cap [0,9] = [0,4],$$
$$\text{but } f(A_1 \cap A_2) = f(\{0\}) = \{0\}.$$

(b) (i)

$$x \in \text{LHS} \Leftrightarrow f(x) \in \bigcup_{i \in I} B_i$$
$$\Leftrightarrow \exists i \in I \text{ such that } f(x) \in B_i$$
$$\Leftrightarrow \exists i \in I \text{ such that } x \in f^{-1}(B_i)$$
$$\Leftrightarrow x \in \bigcup_{i \in I} f^{-1}(B_i) = \text{RHS.}$$

(ii) Similar.

(iii)

$$x \in \text{LHS} \Leftrightarrow f(x) \in Y \setminus B$$
$$\Leftrightarrow f(x) \notin B$$
$$\Leftrightarrow x \notin f^{-1}(B)$$
$$\Leftrightarrow x \in X \setminus f^{-1}(B) = \text{RHS.}$$

(c) (i)

$$y \in \text{LHS} \Rightarrow \exists x \in f^{-1}(B) \text{ such that } y = f(x)$$
$$\Rightarrow y \in B = \text{RHS.}$$

(ii)

$$x \in A \Rightarrow f(x) \in f(A)$$
$$\Rightarrow x \in f^{-1}(f(A)) = \text{RHS.}$$

(iii) (A) Suppose f is onto.

For any $y \in B, \exists x \in X$ such that $y = f(x)$

and then $x \in f^{-1}(B)$,
so $y \in f(f^{-1}(B))$.
This shows $B \subseteq f(f^{-1}(B))$,
and now (i) shows we have equality of these sets.

(B) Suppose f is not onto.
Then $\exists y \in Y \setminus f(X)$.
Put $B = \{y\}$. Then $f(f^{-1}(B)) = f(\emptyset) = \emptyset \neq B$.

(iv) (A) Suppose f is 1–1.
For any $x \in f^{-1}(f(A)), f(x) \in f(A)$,
so $\exists x' \in A$ such that $f(x) = f(x')$.
But (since f is 1–1) $x = x'$,
so $x \in A$.
This shows $f^{-1}(f(A)) \subseteq A$,
and now (ii) shows we have equality of these sets.

(B) Suppose f is not 1–1.
Choose $x_1 \neq x_2$ such that $f(x_1) = f(x_2)$.
Put $A = \{x_1\}$. Then $x_2 \in f^{-1}(f(A))$,
so $A \neq f^{-1}(f(A))$.

1.7 Suppose, if possible, that:

$C \subseteq [0, \omega_1)$, C is countable and C is *not* bounded above in $[0, \omega_1)$.

For each $x \in [0, \omega_1)$, x is *not* an upper bound of C,
so $x < t_x$ for some $t_x \in C$.
That is, $[0, \omega_1) \subseteq \bigcup_{t \in C}[0, t)$.
But each $[0, t)$ (where $t < \omega_1$) is countable,
so $\bigcup_{t \in C}[0, t)$ is a countable union of countable sets, and therefore countable.
This is a contradiction, since $[0, \omega_1)$ is uncountable.

1.11 (i) Using arithmetic in base 5, the typical element of *Cantor*$_5$ can be represented as a 'pentimal' $0.a_1a_2a_3a_4 \cdots_{[5]}$, where each base-5 digit a_i is either 0 or 4.
The average of two of these (a and b, say) will therefore take the form $0.m_1m_2m_3m_4 \cdots$, where each m_i is either 0 or 2 or 4.
If a and b are distinct, one of them will have a 0 in a pentimal place where the other has a 4, so at least one m_i will equal 2.

Then it will not be possible to represent $(a + b)/2$ as a pentimal using only 0s and 4s as digits,
so $(a + b)/2$ does not belong to $Cantor_5$.

(ii) If we try the same argument in base 3 to describe $Cantor_3$, then 0 and 2 are the permitted digits and the conclusion reached is that at least one digit m_i of the average of a and b equals 1.
But it may still be possible to represent that average without using the digit 1 because, for instance, $0.02201_{[3]}$ and $0.02200\dot{2}_{[3]}$ are the same base-3 number;
hence the argument of (i) above will fail.
More explicitly, 0 and 2/3 belong to $Cantor_3$ (in tresimal form they are 0.0 and 0.2) but so is their average 1/3, since it can be expressed not only as 0.1 but also as $0.0\dot{2}$.

Topological spaces

Some elementary concepts

2.1 Definition Let X be a non-empty set. A *topology* on X is a collection τ of subsets of X such that

(i) X and $\emptyset$ belong to τ,

(ii) the union of **any** set of members of τ is a member of τ,

(iii) the intersection of **any finite** set of members of τ is a member of τ.

The pair of objects (X, τ) is called a *topological space*. We often write it simply as X, provided that it is obvious *which* topology we have in mind. The members of τ are called the *open sets*, or the *τ-open sets*. Due, perhaps, to the geometrical and metric-space roots of the subject, the elements of X are often referred to as the *points* of the space.

2.2 Examples

(i) For any metric space (M, d), we know that its open sets do form a topology on M, which we shall usually write as τ_d. Any topology that arises like this is said to be *metrisable*. In particular, the usual metrics on $\mathbb{R}^n$ (for each $n \geq 1$) and on its subsets give rise to topologies on $\mathbb{R}$ and its powers and their various subsets, which we refer to as the *usual topologies* on these sets. The notation τ_{usual} will be used to denote these 'most familiar' topologies.

(ii) On any set X, $\{X, \emptyset\}$ is a (not very interesting!) topology, called the *trivial topology*, which we sometimes denote by τ_{triv}.

(iii) At the other extreme, any set X supports a so-called *discrete topology* τ_{disc}, defined to be the entire powerset $P(X)$ (that is, the set of *all* subsets of X). This happens to be metrisable: the metric which assigns a distance of 1 to each two distinct elements of X produces it.

(iv) If X is infinite, then a subset A of X is called *cofinite* when its complement $X \setminus A$ is finite. The collection

$$\tau_{\text{cf}} = \{G \subseteq X : G \text{ is cofinite or } G = \emptyset\}$$

is a topology on X, the *cofinite topology* on this set.

(v) If X is uncountable, then a subset A of X is called *cocountable* when its complement $X \setminus A$ is countable. The collection

$$\tau_{cc} = \{G \subseteq X : G \text{ is cocountable or } G = \emptyset\}$$

is a topology on X, the *cocountable topology* on this set. (You should check that it is a topology.)

(vi) If X is any set and p is an arbitrary point of X, then the collection

$$\iota_p = \{G \subseteq X : p \in G \text{ or } G = \emptyset\}$$

is a topology on X, the *included-point topology (based at p)* there.

(vii) If X is any set and p is an arbitrary point of X, then the collection

$$\epsilon_p = \{G \subseteq X : p \in X \setminus G \text{ or } G = X\}$$

is a topology on X, the *excluded-point topology (based at p)* there.

We shall feel free to refer to a discrete space, a trivial space, a cofinite space and so on, when the topology in play is so named. Needless to say, there are many examples of much more interesting topologies than these, and we shall see some of them. However, the above handful of basic examples will turn out to be surprisingly useful.

2.3 Definition Let (X, τ) be a (topological) space, p an element of X and N a subset of X. We call N a *neighbourhood* of p (or, when we need to be fussy, a *τ-neighbourhood* of p) if there is an open set G in τ for which

$$p \in G \subseteq N$$

(Fig. 2.1)

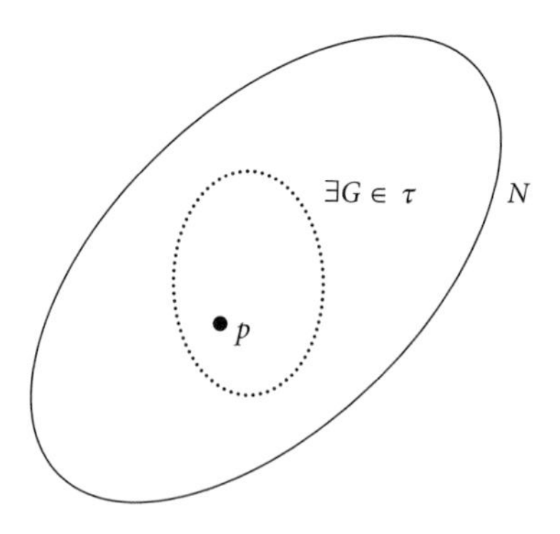

Fig. 2.1 N is a τ-neighbourhood of p (see 2.3).

2.4 Exercise Determine how to recognise neighbourhoods of points in the spaces of 2.2. Keep in mind that neighbourhoods do not have to be open sets themselves.

2.5 Lemma The intersection of finitely many neighbourhoods of a point p (in any topological space) is a neighbourhood of p.

2.6 Lemma Given a space (X, τ) and any subset A of X, we have that $A \in \tau$ if and only if A is a τ-neighbourhood of **every** point in A.

2.7 Definition A subset F of a space (X, τ) is called *closed* (or τ-*closed*) when its complement $X \setminus F$ is open.

2.8 Warning Some subsets are both open and closed. Many subsets are neither open nor closed. Therefore:

- to prove that a set is open, it is not enough to show that it cannot be closed;
- to prove that a set is closed, it is not enough to show that it cannot be open;
- to prove that a set is not open, it is not enough to show that it is closed;
- to prove that a set is not closed, it is not enough to show that it is open.

This paragraph highlights the most common source of mistakes in elementary topology!

Because of how we have defined 'closed', it is hardly surprising that the basic properties of the closed sets are mirror images of those of the open sets:

2.9 Proposition In any topological space (X, τ):

(i) X and $\emptyset$ are τ-closed,

(ii) the intersection of **any** set of closed sets is a closed set,

(iii) the union of **any finite** set of closed sets is a closed set.

2.10 Definition Given a subset A of a space (X, τ), the *closure* of A, usually written $\overline{A}$ or, when we really have to, $\overline{A}^{\tau}$, is the intersection of all the τ-closed sets that happen to contain A. It is closed. It is the smallest closed set that does contain A. The set A is closed if and only if $A = \overline{A}$.

Here are the basic properties of the closure process.

2.11 Proposition For any subsets A, B of a space (X, τ):

(i) $\overline{\emptyset} = \emptyset$,

(ii) $A \subseteq \overline{A}$ always,

(iii) $\overline{\overline{A}} = \overline{A}$ always,

(iv) $\overline{A \cup B} = \overline{A} \cup \overline{B}$.

2.12 Warning Note that 2.11(iv) extends easily to the union of any finite number of sets, but that it does not extend to unions of infinitely many.

2.13 Exercise Explore examples of closure in some spaces. For instance, determine the closure of an arbitrary set in a cofinite space, or in an included-point space. Note in particular that, in $(\mathbb{R}, \tau_{\text{usual}})$, $\overline{\mathbb{Q}} = \mathbb{R}$, where $\mathbb{Q}$ is the set of rational numbers, and that the Cantor excluded-middle set (using any arithmetical base) is closed.

Here is a highly useful little result tying together closure and neighbourhoods in any given space.

2.14 Lemma For $A \subseteq X$ and $p \in X$, we have that $p \in \overline{A}$ if and only if **every** neighbourhood of p meets A (Fig. 2.2).

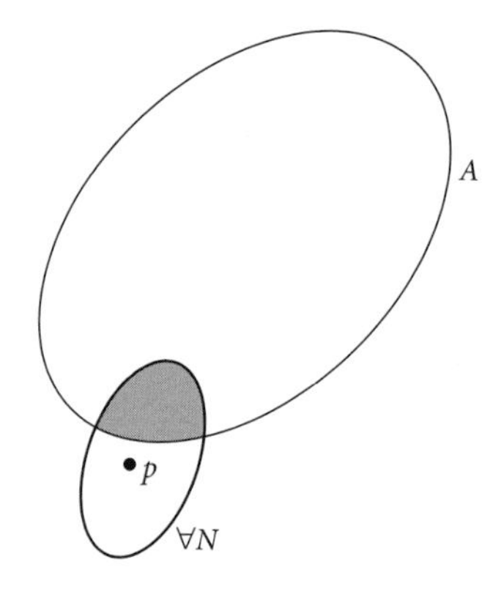

Fig. 2.2 All neighbourhoods of p meet A (see 2.14).

Subspaces

2.15 Definition Let A be a non-empty subset of X, where (X, τ) is a given space. The set of traces onto A of the τ-open sets – by which we mean the collection

$$\{A \cap G : G \in \tau\}$$

– is a topology on A, sometimes written as τ_A or as $\tau|A$. It is called the topology *induced* on A by τ, or, more usually, the *subspace topology*. The space (A, τ_A) is termed a *subspace* of (X, τ) (Fig. 2.3).

2.16 Lemma In the above notation, the τ_A-closed sets are precisely those of the form $A \cap F$, where F is τ-closed.

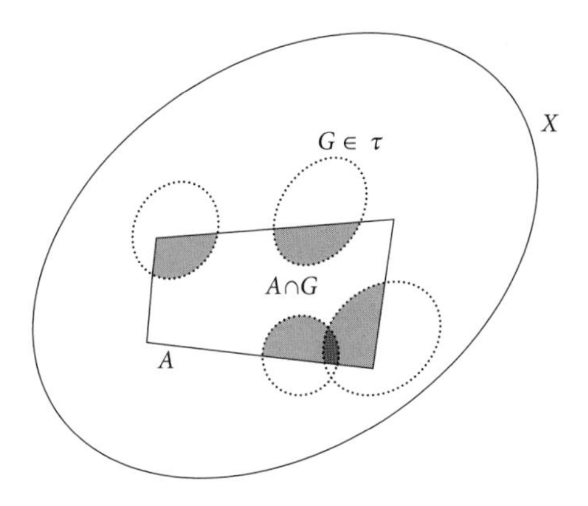

Fig. 2.3 A as a subspace of (X, τ) (see 2.15).

2.17 Lemma In the above notation, given $p \in A$, the τ_A-neighbourhoods of p are precisely the sets of the form $A \cap N$, where N is a τ-neighbourhood of p (Fig. 2.4).

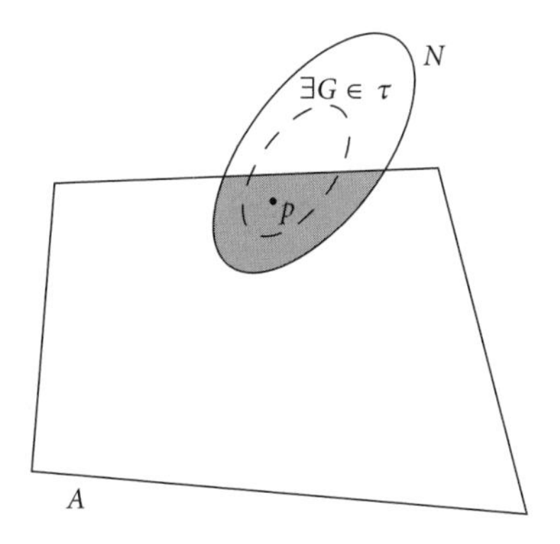

Fig. 2.4 Neighbourhoods in the subspace topology (see 2.17).

2.18 Lemma In the above notation, given $B \subseteq A$, the τ_A-closure of B is

$$\overline{B}^{\tau_A} = A \cap \overline{B}^{\tau}.$$

2.19 Warning Although 2.16–2.18 suggest (correctly) that subspaces are usually easy to handle because the 'structure' just gets traced or shadowed onto the subset that carries the subspace topology, there is also a rich source of errors here. Suppose that we are given (X, τ) and $B \subseteq A \subseteq X$. If we say '$B$ is open', then there are two distinct things that we might mean: that B is a τ-open subset of X which happens to be contained in A, or that B is a τ_A-open set. It is vital to realise that these are different; similar remarks apply to 'closed', to 'neighbourhood', to 'closure' and so on. (You should be able to find easy illustrations of these comments in the real line, using sets no more complicated than intervals.)

A way to prevent this difficulty arising is consistently to use terms like τ-open, τ_A-open, τ-neighbourhood, τ_A-closure and so on whenever there is a genuine risk of ambiguity. We note two small positive results also:

2.20 Lemma If A is itself τ-open, then any τ_A-open set will be also τ-open; if A is itself τ-closed, then any τ_A-closed set will be also τ-closed.

2.21 Definition A property of topological spaces is called *hereditary* if, whenever a space possesses the property, then so must all of its subspaces. An example is metrisability.

Many properties are inherited not by all subspaces but by important special classes of subspaces, and it is useful to extend the definition. A property that is inherited by every subspace (A, τ_A) **for which A is τ-closed** is called a *closed-hereditary* property. A property that is inherited by every subspace (A, τ_A) **for which A is τ-open** is called an *open-hereditary* property.

2.22 Examples The property 'every non-empty open set is uncountable' is not hereditary, but it is open-hereditary. The property 'every countable set is closed' is hereditary, and therefore also closed-hereditary and open-hereditary. Other somewhat contrived examples like these can be devised easily enough but, once again, we shall see better and more useful examples later.

Exercises Essential Exercises 4–13, 15 and 17 are based on the material in this chapter. It is particularly recommended that you should try numbers 4, 5, 6, 7, 8, 11 and 13.

Expansion of Chapter 2

2.4

(i) N is a neighbourhood of $x \Leftrightarrow \exists \varepsilon > 0$ so that $B(x, \varepsilon) \subseteq N$.

(ii) N is a neighbourhood of $x \Leftrightarrow N = X$.

(iii) N is a neighbourhood of $x \Leftrightarrow x \in N$.

(iv) N is a neighbourhood of $x \Leftrightarrow x \in N$ and $X \setminus N$ is finite.

(v) N is a neighbourhood of $x \Leftrightarrow x \in N$ and $X \setminus N$ is countable.

(vi) N is a neighbourhood of $x \Leftrightarrow p \in N$ and $x \in N$.

(vii) N is a neighbourhood of $x \Leftrightarrow$

$$\begin{cases} x = p \quad \& N = X \\ \text{or} \\ x \neq p \quad \& \text{ either } (x \in N \text{ and } p \notin N) \text{ or } N = X. \end{cases}$$

2.5 Let $N_1, N_2, \ldots, N_j$ be finitely many neighbourhoods of p (in X).
Choose open sets $G_1, G_2, \ldots, G_j$ so that

$$p \in G_1 \subseteq N_1, \quad p \in G_2 \subseteq N_2, \ldots, p \in G_j \subseteq N_j.$$

Then $\bigcap_1^j G_i$ is open, and $p \in \bigcap_1^j G_i \subseteq \bigcap_1^j N_i$

therefore $\bigcap_1^j N_i$ is a neighbourhood of p.

2.6

(i) Suppose A is τ-open in (X, τ).
For each $x \in A$, we choose $G = A \in \tau$ and we have $x \in G \subseteq A$.
Therefore A is a neighbourhood of x.
That is, A is a neighbourhood of every point in A.

(ii) Suppose A is a neighbourhood of each of its own points. So:
for each $x \in A$ we can choose $G_x \in \tau$ such that $x \in G_x \subseteq A$.
Now $\bigcup_{x \in A} G_x \subseteq A$ obviously. But also each point x of A belongs to G_x, so $A \subseteq \bigcup_{x \in A} G_x$.
Thus $A = \bigcup_{x \in A} G_x$, a union of τ-open sets, therefore τ-open itself.

2.9

(i) X and $\emptyset$ are closed because their complements ($\emptyset$ and X) are open.

(ii) If $\{F_i : i \in I\}$ is any family of closed sets, then *each* $X \setminus F_i$ is open, so $\bigcup_{i \in I}(X \setminus F_i)$ is open.

$$\text{But} \quad X \setminus \bigcap_{i \in I} F_i = \bigcup_{i \in I} (X \setminus F_i)$$

therefore $\bigcap_i F_i$ is closed.

(iii) If $F_1, F_2, \ldots, F_n$ are finitely many closed sets then

$$X \setminus \bigcup_1^n F_i = \bigcap_1^n (X \setminus F_i)$$

= a finite intersection of open sets therefore open,

so $\bigcup_1^n F_i$ is closed.

2.11

(iv) $\overline{A}_1 \cup \overline{A}_2$ is closed and contains $A_1 \cup A_2$

$$\begin{aligned} \text{therefore} \quad \overline{A_1 \cup A_2} &\subseteq \overline{A}_1 \cup \overline{A}_2. & (1) \\ \text{But also } \textbf{each } A_i &\subseteq A_1 \cup A_2, \\ \text{therefore} \quad \overline{A}_i &\subseteq \overline{A_1 \cup A_2} \\ \text{therefore} \quad \overline{A}_1 \cup \overline{A}_2 &\subseteq \overline{A_1 \cup A_2}. & (2) \end{aligned}$$

By (1) and (2) we have

$$\overline{A_1 \cup A_2} = \overline{A}_1 \cup \overline{A}_2.$$

2.12 $\overline{A}_1 \cup \overline{A}_2 \cup \ldots \cup \overline{A}_n$ is closed and contains $A_1 \cup A_2 \cup \ldots \cup A_n$.

$$\begin{aligned} \text{Therefore} \quad \overline{A_1 \cup A_2 \cup \ldots \cup A_n} &\subseteq \overline{A}_1 \cup \overline{A}_2 \cup \ldots \cup \overline{A}_n. & (1) \\ \text{But also } \textbf{each } A_i &\subseteq A_1 \cup \ldots \cup A_n \\ \text{therefore} \quad \overline{A}_i &\subseteq \overline{A_1 \cup \ldots \cup A_n} \\ \text{therefore} \quad \bigcup_1^n \overline{A}_i &\subseteq \overline{A_1 \cup \ldots \cup A_n}. & (2) \end{aligned}$$

By (1) and (2) we have

$$\overline{\bigcup_1^n A_i} = \bigcup_1^n \overline{A}_i.$$

In $(\mathbb{R}, \tau_{\text{usual}})$:

$$\overline{\bigcup_1^\infty \left(\frac{1}{n+1}, 1\right)} = \overline{(0,1)} = [0,1],$$

$$\text{but} \quad \bigcup_1^\infty \overline{\left(\frac{1}{n+1}, 1\right)} = \bigcup_1^\infty \left[\frac{1}{n+1}, 1\right] = (0,1].$$

2.13 If the complement in $\mathbb{R}$ of the closure of $\mathbb{Q}$ were non-empty, it would be open and would contain an open interval (a, b). This interval would contain no rational numbers, contradicting the well-known 'density' of the rationals in the reals.

In the real line, each bounded closed interval $[a, b]$ is a closed set because its complement is the union of two (unbounded) open intervals, so the sets by which we recursively constructed the Cantor set – finite unions of bounded closed intervals – are closed. The Cantor set was defined as the intersection of these, and is thus closed also.

2.14

(i) Let $p \in \overline{A}$.
Suppose that N is any neighbourhood of p.
If N does not meet A (that is, their intersection is empty) then choose $G \in \tau$ such that

$$p \in G \subseteq N$$

and observe that G does not meet A either,

$$\text{that is,} \quad X \setminus G \supseteq A, \quad \text{where } X \setminus G \text{ is closed,}$$
$$\text{therefore} \quad X \setminus G \supseteq \overline{A}.$$

Now $p \in \overline{A}$ tells us $p \in X \setminus G$, that is, $p \notin G$, *contradiction!*
So N has to intersect A.

(ii) Let $p \notin \overline{A}$.
Then $X \setminus \overline{A}$ is an open set including p
and therefore is a neighbourhood of p
and it certainly cannot intersect A!
So *not every* neighbourhood of p meets A.

2.16 Let $F_1 \subseteq A \subseteq X$.

(i) Suppose F_1 is τ_A-closed
that is, $A \setminus F_1$ is τ_A-open
that is, $A \setminus F_1 = A \cap G$ for some $G \in \tau$.
But then $F_1 = A \cap (X \setminus G)$
and $X \setminus G$ is τ-closed.

(ii) Suppose F_1 takes the form $A \cap F$, where F is τ-closed.
Then $X \setminus F$ is τ-open
and $A \cap (X \setminus F) = A \setminus F_1$ is τ_A-open.
Therefore F_1 is τ_A-closed.

2.17 Let $p \in A \subseteq X, N_1 \subseteq A$.

(i) Suppose N_1 is a τ_A-neighbourhood of p.
Then $\exists G_1 \in \tau_A$ such that $p \in G_1 \subseteq N_1$
and G_1 takes the form $A \cap G$, some $G \in \tau$.
Now $p \in G \subseteq N_1 \cup G$
so $N_1 \cup G$ is a τ-neighbourhood of p
and

$$A \cap (N_1 \cup G) = (A \cap N_1) \cup (A \cap G) = N_1 \cup G_1 = N_1,$$

so N_1 is of the form $A \cap$ 'a τ-neighbourhood of p'.

(ii) Suppose $N_1 = A \cap N$, where N is a τ-neighbourhood of p,
that is, $p \in G \subseteq N$ for some $G \in \tau$.
Then $p \in A \cap G \subseteq A \cap N$,
that is, $p \in A \cap G \subseteq N_1$,
where $A \cap G$ *is* τ_A-open.
Therefore N_1 is a τ_A-neighbourhood of p.

2.18 Let $B \subseteq A \subseteq X$.
Now $A \cap \overline{B}^{\tau}$ is τ_A-closed (see 2.16)
and contains $A \cap B = B$.

$$\text{Therefore } \overline{B}^{\tau_A} \subseteq A \cap \overline{B}^{\tau}. \tag{1}$$

On the other hand, any τ_A-closed set that contains B takes the form $A \cap F$, where F is τ-closed

$$\begin{aligned} \text{and} \quad & B \subseteq F \\ \text{therefore} \quad & \overline{B}^{\tau} \subseteq F \\ \text{therefore} \quad & A \cap \overline{B}^{\tau} \subseteq A \cap F. \end{aligned}$$

Yet $\overline{B}^{\tau_A}$ is the intersection of all such sets $A \cap F$,

$$\text{therefore} \quad A \cap \overline{B}^{\tau} \subseteq \overline{B}^{\tau_A}. \tag{2}$$

Now we combine (1) and (2).

2.19 Let $X = (\mathbb{R}, \tau_{\text{usual}}), A = (0, 1]$.
Then $(0, \frac{1}{2}]$ is closed in the subspace A
(since it equals $A \cap [0, 1]$)
but not closed in X.
Also, $(\frac{1}{2}, 1]$ is open in the subspace A
(since it equals $A \cap (\frac{1}{2}, \frac{3}{2})$)
but not open in X.
Also, $(\frac{1}{2}, 1]$ is a subspace-neighbourhood of 1
(since it equals $A \cap (\frac{1}{2}, \frac{3}{2})$)
but is not a neighbourhood of 1 in X.

2.20 Let A be τ-open in X, B ($\subseteq A$) be τ_A-open.
Then $B = A \cap G$ for some $G \in \tau$
the intersection of two τ-open sets,
therefore τ-open.
Let A be τ-closed in X, B ($\subseteq A$) be τ_A-closed.
Then $B = A \cap F$ for some τ-closed F
the intersection of two τ-closed sets,
therefore τ-closed.

2.21 Let (A, τ_A) be a subspace of *metrisable* (X, τ).
Choose a metric d on X so that τ_d is τ.
Then $d|_{A\times A}$ is a metric on A. Call it d'.
We claim: $\tau_{d'}$ is τ_A.

Let $p \in A, N \subseteq A$. Then:

N is a $\tau_{d'}$-neighbourhood of p
$\Leftrightarrow \exists \varepsilon > 0$ for which $\{q \in A: d'(p, q) < \varepsilon\} \subseteq N$
$\Leftrightarrow \exists \varepsilon > 0$ such that $A \cap \{q \in X: d(p, q) < \varepsilon\} \subseteq N$
$\Leftrightarrow N$ contains the trace onto A of a τ_d-neighbourhood of p
$\Leftrightarrow N$ *is* the trace onto A of a τ_d-neighbourhood of p
$\Leftrightarrow N$ is a τ_A-neighbourhood of p.

Since $\tau_{d'}$ and τ_A yield the same neighbourhoods of points of A, they are the same topology (see, for example, 2.6).

2.22

- Let (X, τ) satisfy 'every non-empty open set is uncountable'.
Let (A, τ_A) be one of its subspaces, where A is open.
Any τ_A-open (non-empty) set is also τ-open (see 2.20),

therefore uncountable,
so (A, τ_A) possesses the same property.

- $(\mathbb{R}, \tau_{\text{usual}})$ satisfies the condition 'every non-empty open set is uncountable' but its subspace $\mathbb{Z}$ certainly does not!

- Let (X, τ) satisfy 'every countable set is closed'.
Let (A, τ_A) be one of its subspaces.
Any $C \subseteq A$ that is countable
is countable as a subset of X
and is therefore closed in τ.
Now $C = A \cap C$ is also τ_A-closed.
So (A, τ_A) enjoys the same property.

3 Continuity and convergence

Continuity

3.1 Definition A mapping $f : (X, \tau) \to (Y, \tau')$ from a topological space to a topological space is called *continuous* if

$$G \in \tau' \Rightarrow f^{-1}(G) \in \tau;$$

that is, if the inverse image of every open set is open.

There are several different ways to recognise continuity; here are two of them:

3.2 Lemma For a mapping $f : (X, \tau) \to (Y, \tau')$, the following are equivalent:

(i) f is continuous;
(ii) for each τ'-closed subset K of $Y, f^{-1}(K)$ is τ-closed in X;
(iii) for each $A \subseteq X, f(\overline{A}) \subseteq \overline{f(A)}$.

3.3 Examples

(i) A map from a metric space to a metric space is continuous (in the metric sense) if and only if it is continuous between the topological spaces induced by the metrics. Therefore, all the examples of continuity you knew earlier, you still know!
(ii) Any map whose domain is discrete must be continuous. Any map whose codomain is trivial must be continuous.
(iii) Constant maps are continuous.

3.4 Lemma The composite of two continuous maps is continuous.

3.5 Lemma Any restriction of a continuous map is continuous.

3.6 Lemma Given a mapping $f : (X, \tau) \to (Y, \tau')$ and a set Z satisfying $f(X) \subseteq Z \subseteq Y$, let f^* be the same map as f except that its codomain is Z. Then f is continuous if and only if f^* is continuous.

3.7 Examples

(i) Any one-to-one map from $(\mathbb{R}, \tau_{cc})$ to $(\mathbb{R}, \tau_{cf})$ is continuous.

(ii) Any continuous map from $(\mathbb{R}, \tau_{cf})$ to $(\mathbb{R}, \tau_{usual})$ is constant.

3.8 Definition A mapping $f : (X, \tau) \to (Y, \tau')$ is called a *homeomorphism* if

(i) it is one-to-one and onto, and

(ii) it is continuous, and

(iii) the inverse map f^{-1} is continuous.

Whenever such a map exists between two spaces, they are called *homeomorphic*, which really means that they have identically the same behaviour as topological spaces (compare 'isomorphism' in groups or 'linear isomorphism' in vector spaces).

A *topological property* or *homeomorphic invariant* is a statement about topological spaces which, whenever it is true for a space (X, τ), is necessarily true also for any space that is homeomorphic with (X, τ). In effect, these are the properties/statements that can be expressed purely in terms of open sets plus set theory. Two spaces are homeomorphic if and only if they have exactly the same topological properties.

3.9 Examples

(i) The identity map on a space is a homeomorphism. The composite of two homeomorphisms is a homeomorphism. The inverse of a homeomorphism is a homeomorphism. 'Being homeomorphic' is therefore an equivalence relation on any set of spaces.

(ii) Any two open intervals on the real line $(\mathbb{R}, \tau_{usual})$ are homeomorphic to one another, and are homeomorphic to the entire real line.

(iii) $(\mathbb{R}, \tau_{usual})$ and $(\mathbb{R}, \tau_{cf})$ are not homeomorphic.

3.10 Illustration of homeomorphism: the fractal character of the Cantor sets

Notice that the subset $Cantor_{10}$ of the real line that we described in 1.10 cannot contain any interval: for at stage n of the process, the initial interval of length 1 has been reduced to $*^n[0, 1]$, the union of 2^n disjoint subintervals each of length 10^{-n}; therefore if J denotes a non-degenerate interval of positive length δ, we need only choose a positive integer n such that $10^{-n} < \delta$ to deduce that J cannot be contained in $*^n[0, 1]$, let alone in $Cantor_{10}$ itself.

Consider now a typical one of the intervals created at stage n of the iteration: say, $[p, q]$. All the stage-n intervals have the same length 10^{-n}, so $q = p + 10^{-n}$ and the map $x \mapsto x - p$ transforms $[p, q]$ onto $[0, 10^{-n}]$. Furthermore, since p as a decimal consists of an initial block of exactly n-many 0s and 9s, followed by an endless stream of 0s (which, as in conventional arithmetic, we can safely ignore), subtraction of p does not change any decimal digit after the nth. It follows that

$x \mapsto x - p$ actually transforms $Cantor_{10} \cap [p, q]$ onto $Cantor_{10} \cap [0, 10^{-n}]$. By essentially the same argument, $x \mapsto (10^n)x$ transforms $Cantor_{10} \cap [0, 10^{-n}]$ onto $Cantor_{10}$ itself. Since both of these transformations are continuous and continuously reversible, we conclude that the part of $Cantor_{10}$ that falls inside an arbitrary interval created at some stage in the iterative process that we described is homeomorphic to the entire space. The critical point to make is that, if x is any point of $Cantor_{10}$ and N is any $Cantor_{10}$-neighbourhood of x, then we can select $\varepsilon > 0$ so small that $Cantor_{10} \cap (x - \varepsilon, x + \varepsilon)$ lies inside N, and now x must belong to an interval (created at some stage in the process) of length less than ε. Thus $Cantor_{10}$ enjoys one of the characteristic properties of the so-called 'fractals': every neighbourhood of a typical point contains a homeomorphic copy of the whole space. That the same is true for the 'other' Cantor sets is best seen by establishing the following result:

3.11 Proposition Cantor excluded-middle sets arising from different arithmetical bases are homeomorphic.

(*Suggested method*) There is a natural bijection θ from $Cantor_3$ to $Cantor_{10}$: it is the act of replacing 2 by 9 in the tresimal representation of each element of $Cantor_3$ and of interpreting the result as a decimal. To be slightly more formal, it is

$$\theta\left(\sum \frac{a_n}{3^n}\right) = \sum\left(\frac{9a_n/2}{10^n}\right).$$

Since θ merely substitutes 9s for 2s, it preserves the block structures by which we built $Cantor_3$ and $Cantor_{10}$: for instance, if x falls in the 17th block from the left created at stage 11 in our iterative construction for $Cantor_3$, then $\theta(x)$ must belong to the 17th block from the left created at stage 11 when we built $Cantor_{10}$ (and vice versa). Notice the lengths (10^{-n} for $Cantor_{10}$, 3^{-n} for $Cantor_3$) of the subinterval blocks that are created at stage n as we construct towards these fractals, and that the gaps between these blocks are in every case at least as big as the blocks themselves.

Because of Proposition 3.11, it is safe henceforth to refer to *the Cantor set* whenever we intend merely to focus on its properties as a topological space (rather than explicitly as a subset of the real line) without reference to which arithmetical base we use in constructing it.

We now briefly consider two other properties of mappings, which resemble continuity but are less important.

3.12 Definitions

(i) A mapping $f : (X, \tau) \to (Y, \tau')$ is called *open* if

$$G \in \tau \Rightarrow f(G) \in \tau',$$

that is, if the *direct* image of every open set is open.

(ii) A mapping $f : (X, \tau) \to (Y, \tau')$ is called *closed* if

$$K \text{ is } \tau\text{-closed} \Rightarrow f(K) \text{ is } \tau'\text{-closed},$$

that is, if the *direct* image of every closed set is closed.

3.13 Examples

(i) Any map into a discrete space is both an open map and a closed map. Any onto map with trivial domain space is both an open map and a closed map.

(ii) Any onto map from $(\mathbb{R}, \tau_{cf})$ to $(\mathbb{R}, \tau_{cc})$ is a closed map.

3.14 Proposition For a one-to-one and onto mapping $f : (X, \tau) \to (Y, \tau')$, the following are equivalent:

(i) f is continuous,

(ii) f^{-1} is an open map,

(iii) f^{-1} is a closed map.

3.15 Corollary For a one-to-one, onto and continuous mapping $f : (X, \tau) \to (Y, \tau')$, the following are equivalent:

(i) f is a homeomorphism,

(ii) f is an open map,

(iii) f is a closed map.

Convergent sequences

3.16 Definition A sequence $(x_n)_{n\geq 1}$ of points in a topological space (X, τ) is said to *converge* to a *limit* l in X if:

$$\text{for each neighbourhood } U \text{ of } l, \ \exists\, n(U) \in \mathbb{N}$$
$$\text{such that } n \geq n(U) \Rightarrow x_n \in U.$$

We then write $x_n \to l$ (Fig. 3.1).

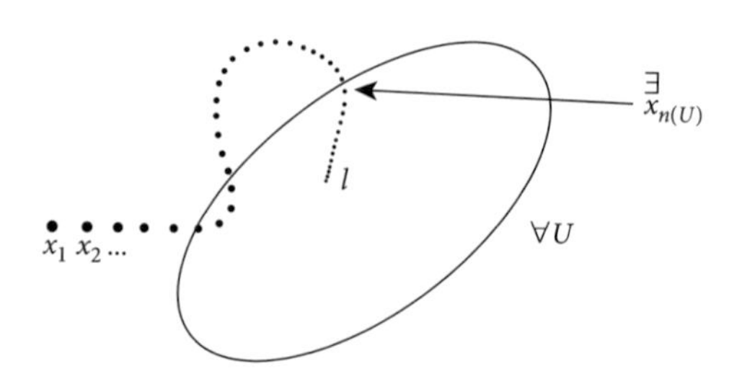

Fig. 3.1 The sequence $(x_n)_{n\geq 1}$ converges to l (see 3.16).

3.17 Examples

(i) This incorporates sequential convergence in the senses used in real analysis, complex analysis and metric space theory. Thus, all the examples you know already are still valid in the topological setting.

(ii) In a trivial space, every sequence converges to every point as a limit! So the uniqueness of limit that we have come to depend upon in metric spaces does not hold good in topological spaces (at least, not in *all* topological spaces).

(iii) In any space, if $(x_n)_{n\geq 1}$ is *eventually constant* (that is, if there is some positive integer n_0 such that for every $n \geq n_0$ we get $x_n = x_{n_0}$), then $(x_n)_{n\geq 1}$ certainly converges.

(iv) In a discrete space, this is the only way a sequence can converge.

(v) In a cocountable space, this is again the only way a sequence can converge.

(vi) Think about what (iv) and (v) tell us: that convergent sequences cannot spot the difference between, say, $(\mathbb{R}, \tau_{cc})$ and $(\mathbb{R}, \tau_{disc})$, even though they are 'obviously' very different spaces.

Powerful though they are in analysis and in metric spaces, convergent sequences already look to be less effective in coping with the structure of general topological spaces. The theme continues below.

3.18 Proposition If G is open in a space (X, τ) then:

$$\text{whenever } x \in G \text{ and } x_n \to x, \text{ we get } x_n \in G \text{ for all sufficiently large } n. \qquad (1)$$

3.19 Proposition If, in (X, τ), there is a sequence of elements all belonging to the subset A and converging to $l \in X$, then $l \in \overline{A}$.

3.20 Proposition If the mapping $f : (X, \tau) \to (Y, \tau')$ is continuous, then:

$$\text{whenever } x_n \to l \text{ in } X, \text{ then } f(x_n) \to f(l) \text{ in } Y. \qquad (2)$$

You probably know that in real/complex analysis and in metric spaces, the converses of 3.18, 3.19 and 3.20 are true as well. The main point to make now is that, in general topological spaces, all three converses are false.

3.21 Example For instance, in $(\mathbb{R}, \tau_{cc})$, the converses of the last three propositions fail.

Fortunately, it is not in metric spaces alone that sequences fully describe what is happening topologically. We next identify a broader category of spaces in which they are just as effective.

3.22 Definition Let (X, τ) be a space and p an element of X. A *countable local base at* p is a sequence

$$N_1, N_2, N_3, \ldots$$

of neighbourhoods of p such that

(i) $N_1 \supseteq N_2 \supseteq N_3 \supseteq \ldots$ and

(ii) every neighbourhood of p contains one of the N_i (Fig. 3.2).

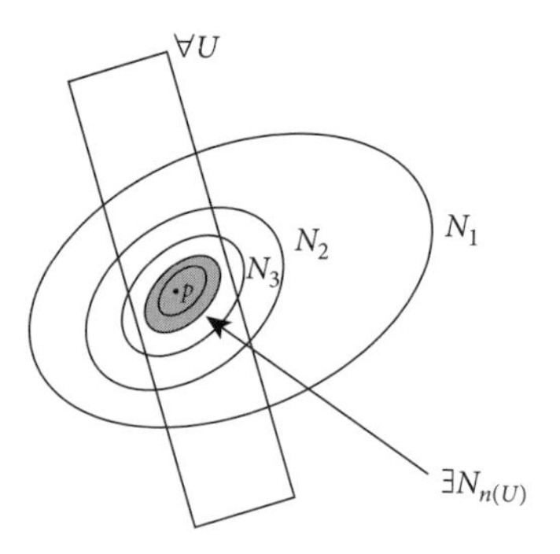

Fig. 3.2 $(N_1, N_2, N_3, \ldots)$ is a countable local base at p (see 3.22).

3.23 Example In any metric space, at any point p, the open balls $B(p, 1/n)$ for $n \geq 1$ form a countable local base.

3.24 Definition A topological space (X, τ) is called *first-countable* if, at each of its points, there is a countable local base.

For instance, any metrisable space is first-countable. Now we shall make the point that sequences are 'good' for first-countable spaces, which is the underlying reason why they are 'good' for metric spaces.

3.25 Proposition A subset G of a first-countable space (X, τ) is open if and only if:

$$\text{whenever } x \in G \text{ and } x_n \to x, \text{ we get } x_n \in G \text{ for all sufficiently large } n. \tag{1}$$

3.26 Proposition If, in a first-countable space (X, τ), p is an element and A is a subset, then $p \in \overline{A}$ if and only if there is a sequence of elements all belonging to A that converges to p.

3.27 Proposition Suppose that there is given a mapping $f : (X, \tau) \to (Y, \tau')$, where (Y, τ') is first-countable. Then f is continuous if and only if:

$$\text{whenever } x_n \to l \text{ in } X, \text{ then } f(x_n) \to f(l) \text{ in } Y. \tag{2}$$

Nets

There is a way to get round the failure of sequences fully to describe (non-first-countable) topological spaces, and it amounts to changing the definition of 'sequence' so as to allow the underlying ordered set – which for sequences is always the positive integers, naturally ordered – to be bigger and/or more complicated.

3.28 Definition A *directed set* $(D, \leq)$ is a non-empty set D together with a *quasiorder* $\leq$ in which each two elements possess a common upper bound. That is, $\leq$ is reflexive and transitive, and for every $a, b \in D$ there is $u \in D$ such that both $a \leq u$ and $b \leq u$.

3.29 Examples The positive integers with their natural (total) order form a directed set. Any ordinal is a directed set. The positive integers under divisibility form a directed set (for instance, the LCM of a and b will serve here as an upper bound for a and b). Most importantly, if X is any given topological space, x is any given point of X and $\mathcal{N}(x)$ is the set of all neighbourhoods of x in X, and we choose to order $\mathcal{N}(x)$ by **inverse inclusion** – that is, $N_1 \leq N_2$ means $N_2 \subseteq N_1$ – then $\mathcal{N}(x)$ becomes a directed set: for any two neighbourhoods of x, their *intersection* will be a common upper bound.

3.30 Definition A *net* in a topological space X is a map from some directed set D into X. We normally denote a typical net not by a function-style notation such as $x : D \to X$ but by a notation such as $(x_\gamma)_{\gamma \in D}$, partly to bring out the (many) parallels between this notion and that of a sequence in X.

3.31 Definition A net $(x_\gamma)_{\gamma \in D}$ of points in a topological space (X, τ) is said to *converge* to a *limit* l in X if:

$$\text{for each neighbourhood } U \text{ of } l, \ \exists \gamma(U) \in D$$
$$\text{such that } \gamma \geq \gamma(U) \Rightarrow x_\gamma \in U.$$

We then write $x_\gamma \to l$.

(We often use the informal phrase 'for all sufficiently large γ' to mean 'for all γ that are $\geq$ some threshold value $\gamma(U)$ in the underlying directed set'. Thus the condition in 3.31 can be verbalised as 'each neighbourhood of l contains x_γ for all sufficiently large values of γ.')

It should be obvious that net convergence incorporates sequence convergence as we described it earlier, but it goes much further by allowing us to adjust the underlying set of labels – no longer constrained to be merely $\mathbb{N}$ all the time! – so as to cope with points whose neighbourhood structures are complicated. The following makes that point more formally:

3.32 Lemma Given a point x in a topological space (X, τ), and using the symbol $\mathcal{N}(x)$ to stand for the family of all neighbourhoods of x in X, suppose that we arbitrarily select from each N in $\mathcal{N}(x)$ an element $x_N \in N$. Then the net $(x_N)_{N \in \mathcal{N}(x)}$ converges to x.

It is absolutely routine to check that 3.18, 3.19 and 3.20 remain true if we replace 'sequence' by 'net' throughout. Much more interestingly, their converses become true when we use nets in place of sequences; that is to say:

3.33 Proposition If G is a subset of a space (X, τ) such that

$$\text{whenever } x \in G \text{ and a net } x_\gamma \to x, \text{ we get } x_\gamma \in G \text{ for all sufficiently large } \gamma, \tag{1}$$

then G is open.

3.34 Proposition If, in a space (X, τ), $p \in \overline{A}$, then there is a net of elements all belonging to the subset A that converges to p.

3.35 Proposition If a mapping $f : (X, \tau) \to (Y, \tau')$ satisfies the condition

$$\text{whenever a net } x_\gamma \to l \text{ in } X, \text{ then } f(x_\gamma) \to f(l) \text{ in } Y, \tag{2}$$

then f is continuous.

Therefore the nice thing about these converses also being valid is that nets fully describe what goes on in topological spaces just as sequences do in metric spaces and in first-countable spaces; that is:

3.36 Proposition A subset G of a topological space (X, τ) is open if and only if:

$$\text{whenever } x \in G \text{ and a net } x_\gamma \to x, \text{ we get } x_\gamma \in G \text{ for all sufficiently large } \gamma. \tag{1}$$

3.37 Proposition If, in a space (X, τ), p is an element and A is a subset, then $p \in \overline{A}$ if and only if there is a net of elements all belonging to A that converges to p.

3.38 Proposition Consider a mapping $f : (X, \tau) \to (Y, \tau')$. Then f is continuous if and only if:

$$\text{whenever } x_\gamma \to l \text{ in } X, \text{ then } f(x_\gamma) \to f(l) \text{ in } Y. \tag{2}$$

Filters

By this stage, we hope to have persuaded the reader that *nets* provide a natural and fairly intuitive way of importing the notion of *convergence to a limit* into general topological spaces in a manner that is as effective in describing them as sequential convergence is in describing metric spaces. However, it is not the only way. A widely used alternative is the idea of a *filter*. Nets and filters do very much the same job in general topology, and it seems best to concentrate on only one of them when studying the discipline for the first time. However, the student who intends to go further in topology or its applications will encounter filters sooner or later, and so a very brief introduction to them at this point may ultimately pay dividends.

3.39 Definition A non-empty family Λ of subsets of a non-empty set X is called a *filter on* X if it satisfies the following three conditions:

(i) $\emptyset \notin \Lambda$,

(ii) $A, B \in \Lambda \Rightarrow A \cap B \in \Lambda$,

(iii) $(A \in \Lambda, A \subseteq S \subseteq X) \Rightarrow S \in \Lambda$.

3.40 Examples

(i) The collection of all supersets of any chosen non-empty $A \subseteq X$ is a filter on X. The family $\mathcal{N}(x)$ of neighbourhoods of a chosen point x in a topological space (X, τ) is a filter on X.

(ii) The complements of the bounded subsets of the coordinate plane (or, indeed, of any given unbounded metric space) comprise a filter on it. The subsets D of $\mathbb{R}$ for which there exists $a \in \mathbb{R}$ such that $D \supseteq (-\infty, a]$ comprise a filter on $\mathbb{R}$.

(iii) The collection of cofinite subsets of an infinite set X (that is, $\tau_{cf} \setminus \{\emptyset\}$) is a filter on X. Likewise, $\tau_{cc} \setminus \{\emptyset\}$ is a filter on any uncountable set.

Filters on the same set may be compared by simple set inclusion and, in particular, a filter on a topological space may be compared with the filter of neighbourhoods of a point:

3.41 Definition

(i) If Λ, Λ' are filters on a set X such that $\Lambda \subseteq \Lambda'$, we call Λ' a *refinement* of Λ, and say that Λ' is *finer* than Λ.

(ii) A filter Λ on a topological space (X, τ) is said to *converge* to a *limit* $p \in X$ if $\Lambda \supseteq \mathcal{N}(p)$; equivalently, if every neighbourhood of p contains a member of the filter Λ.

Here are two indicators of how filter convergence can be used to describe the behaviour of a topological space:

3.42 Proposition An element p of a topological space (X, τ) lies in the closure of a subset A of X if and only if there is a filter Λ on X such that $A \in \Lambda$ and Λ converges to p.

3.43 Proposition A topological space (X, τ) is compact if and only if every filter on X can be refined to a convergent filter (on X).

3.44 Note Some insight into the relationship between filters and nets on a topological space (X, τ) is offered here.

(i) Given a net $x = (x_\gamma)_{\gamma \in D}$ (where D is, as usual, a directed set) and $\gamma_0 \in D$, by the $\gamma_0{}^{\text{th}}$ *tail of the net* we mean the set $\{x_\gamma : \gamma \geq \gamma_0\}$. Define $\Lambda(x)$ to consist of all subsets of X that contain a tail of the net x. Then $\Lambda(x)$ is a filter on X. Furthermore, $\Lambda(x)$ converges to a limit p if and only if the net x converges to p.

(ii) Given a filter Λ on X, notice that Λ is already a directed set under inverse set inclusion. For each $A \in \Lambda$, choose an element x_A arbitrarily in A. Then $(x_A)_{A \in \Lambda}$ is a net in X. Furthermore, Λ converges to a limit p if and only if every such net $(x_A)_{A \in \Lambda}$ converges to p.

Corresponding approximately to the way in which we used 'countable local base' to extract the essential structure of a system of neighbourhoods, it is often convenient to select out 'enough' elements from a filter to be able to describe the entire thing:

3.45 Definitions

(a) A non-empty family Φ of subsets of a non-empty set X is called a *filterbase on X* if it satisfies the following two conditions:

(i) $\emptyset \notin \Phi$,

(ii) $A, B \in \Phi \Rightarrow \exists C \in \Phi$ such that $C \subseteq A \cap B$.

(b) When Φ is a filterbase on X, the family $\{G \subseteq X : \exists F \in \Phi \text{ such that } F \subseteq G\}$ is a filter on X. Then Φ is said to be a base for this filter.

(c) A filterbase (on a topological space) is said to *converge* to a point p if the filter for which it is a base converges to p (equivalently, if every neighbourhood of p contains a member of the filterbase).

One of the (several) reasons why filterbases are used as well as or instead of filters is that the image of a filter under a mapping is quite often not a filter – that is, if $f : X \to Y$ and Λ is a filter on X, then $f(\Lambda) = \{f(A) : A \in \Lambda\}$ may fail to be a filter on Y – whereas this issue does not arise with filterbases: if Λ here is a filterbase, then $f(\Lambda)$ is also a filterbase. This facilitates the following characterisation of continuous maps via filter convergence:

3.46 Proposition Let $f : (X, \tau) \to (Y, \sigma)$ be a mapping between topological spaces. Then f is continuous if and only if, whenever Φ is a filterbase on X converging to a limit $p \in X$, then the filterbase $f(\Phi)$ converges to $f(p)$ in (Y, σ).

Exercises Essential Exercises 14, 16, 18–29 and 47 are based on the material in this chapter. It is particularly recommended that you should try numbers 19, 20, 21, 24, 26, 27 and 29.

Expansion of Chapter 3

3.2 (i) $\Rightarrow$ (ii)

Suppose f continuous.
If $K \subseteq Y$ is τ'-closed
then $Y \setminus K$ is τ'-open.
Therefore $f^{-1}(Y \setminus K)$ is τ-open,
that is, $X \setminus f^{-1}(K)$ is τ-open,
therefore $f^{-1}(K)$ is τ-closed.

(ii) $\Rightarrow$ (iii)

Suppose f^{-1}(each τ'-closed set) is τ-closed.
Given $A \subseteq X$: $\overline{f(A)}$ is τ'-closed,
therefore $f^{-1}(\overline{f(A)})$ is τ-closed.
But $A \subseteq f^{-1}(f(A)) \subseteq f^{-1}\overline{(f(A)}$,
therefore $\overline{A} \subseteq f^{-1}(\overline{f(A)})$,
that is, $f(\overline{A}) \subseteq \overline{f(A)}$.

(iii)$\Rightarrow$ (i)

Suppose $f(\overline{A}) \subseteq \overline{f(A)}$ for all $A \subseteq X$.
Given open $H \subseteq Y$, try $A = X \setminus f^{-1}(H) = f^{-1}(Y \setminus H)$. So:

$$\begin{aligned} f(\overline{f^{-1}(Y \setminus H)}) &\subseteq \overline{f(f^{-1}(Y \setminus H))} \\ &\subseteq \overline{Y \setminus H} = Y \setminus H, \end{aligned}$$

$$\text{that is, } \overline{f^{-1}(Y \setminus H)} \subseteq f^{-1}(Y \setminus H),$$
$$\text{therefore } f^{-1}(Y \setminus H) \text{ is already } \tau\text{-closed},$$
$$\text{that is, } X \setminus f^{-1}(H) \text{ is } \tau\text{-closed},$$
$$\text{that is, } f^{-1}(H) \text{ is } \tau\text{-open}.$$

So f is continuous.

3.4 Let $f : (X, \tau) \to (Y, \tau'), g : (Y, \tau') \to (Z, \tau'')$ both be continuous.
For any τ''-open $J \subseteq Z, g^{-1}(J)$ is τ'-open.
Therefore $f^{-1}(g^{-1}(J))$ is τ-open.
That is, $(g \circ f)^{-1}(J)$ is τ-open.
Therefore $g \circ f : X \to Z$ is continuous.

3.5 Let:

$$f : (X, \tau) \to (Y, \tau') \text{ be continuous}, \emptyset \neq A \subseteq X,$$
$$g = f|_A \text{ be the restriction of } f \text{ to } A.$$

For any τ'-open $H \subseteq Y$,

$$g^{-1}(H) = \{a \in A: f(a) = g(a) \in H\}$$
$$= A \cap f^{-1}(H),$$

which is τ_A-open.
So g is continuous.

3.6 The τ'_Z-open sets take the form $Z \cap H$, where $H \in \tau'$.
Now $(f^*)^{-1}(Z \cap H)$ and $f^{-1}(H)$ are the same set!
So continuity of f^* and continuity of f demand exactly the same thing.

3.7

(i) Let F be τ_{cf}-closed.
Then either $F = \mathbb{R}$ or F is finite.
Since f is 1–1, F finite $\Rightarrow f^{-1}(F)$ finite also, and therefore countable, which implies that $f^{-1}(F)$ is τ_{cc}-closed.
Of course, if $F = \mathbb{R}$ then $f^{-1}(F) = \mathbb{R}$ also (and is τ_{cc}-closed).
By 3.2, f is continuous.

(ii) Let $f: (\mathbb{R}, \tau_{cf}) \to (\mathbb{R}, \tau_{usual})$ be continuous and suppose that it were *not* constant.
Then we can find $x, y \in \mathbb{R}$ such that $f(x) > f(y)$ (Fig. 3.3).

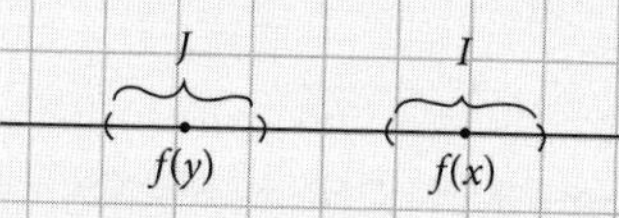

Fig. 3.3 Distinct real numbers lie in disjoint open intervals (see Expansion of 3.7(ii)).

Pick *disjoint* open intervals I, J in $\mathbb{R}$ such that $f(x) \in I$ and $f(y) \in J$.
Now $f^{-1}(I) \cap f^{-1}(J) = f^{-1}(I \cap J) = f^{-1}(\emptyset) = \emptyset$
and $x \in f^{-1}(I)$ and $y \in f^{-1}(J)$,
so $f^{-1}(I)$ and $f^{-1}(J)$ are non-empty, disjoint and τ_{cf}-open. Then

$$\mathbb{R} = \mathbb{R} \setminus \emptyset = \mathbb{R} \setminus (f^{-1}(I) \cap f^{-1}(J))$$
$$= (\mathbb{R} \setminus f^{-1}(I)) \cup (\mathbb{R} \setminus f^{-1}(J))$$
$$= \text{union of two finite sets!} \quad \textit{Contradiction.}$$

3.9

(ii) • For any $a < b$, the formula

$$f(x) = a + (b - a)x \quad (0 < x < 1)$$

gives a homeomorphism from $(0, 1)$ to (a, b).

• For any $a \in \mathbb{R}$, the formula

$$f(x) = \frac{1}{x} - 1 + a \quad (0 < x < 1)$$

gives a homeomorphism from $(0, 1)$ to (a, ∞).

• For any $a, b \in \mathbb{R}$, the formula

$$f(x) = -x + a + b$$

gives a homeomorphism from (a, ∞) to $(-\infty, b)$.

• Also, $f(x) = \tan x \, (-\frac{\pi}{2} < x < \frac{\pi}{2})$ gives a homeomorphism from $(-\frac{\pi}{2}, \frac{\pi}{2})$ to $\mathbb{R}$.

(In each case, look at what the inverse is, to check its continuity!)

Now invoke (i).

3.9

(iii) A one-line proof from 3.7(ii)!

3.11 Take the map θ from $Cantor_3$ to $Cantor_{10}$ suggested in the hint:

$$\theta\left(\sum \frac{a_n}{3^n}\right) = \sum\left(\frac{9a_n/2}{10^n}\right);$$

note that it preserves the block structures by which we constructed the two spaces. We seek a simple proof of continuity for θ:

Given x in $Cantor_3$ and $\varepsilon > 0$, take a value of n such that 10^{-n} is less than ε;

then $y \in Cantor_3$ and $|x - y| < 3^{-n}$ together guarantee that y belongs to the same stage-n block as x,

therefore $\theta(y)$ and $\theta(x)$ fall into the same stage-n block in $Cantor_{10}$'s construction,

whence $|x - y| \leq 10^{-n} < \varepsilon$ and continuity of θ is established.

The argument is reversible, showing that $Cantor_3$ and $Cantor_{10}$ are homeomorphic.
Replace 10 by any other base $b > 3$ and we have that $Cantor_3$ and $Cantor_b$ are homeomorphic.
Hence the result.

3.13

(ii) Let $f\colon (\mathbb{R}, \tau_{cf}) \to (\mathbb{R}, \tau_{cc})$ be onto.
For any closed F in $(\mathbb{R}, \tau_{cf})$:
EITHER F is finite
in which case $f(F)$ is finite,
therefore $f(F)$ is countable,
therefore $f(F)$ is τ_{cc}-closed,
OR $F = \mathbb{R}$
in which case $f(F) = \mathbb{R}$ is again τ_{cc}-closed (remembering that f is onto).

3.14 Because f is 1–1 and onto, $f^{-1}\colon Y \to X$ exists *as a mapping*. So, via definition and 3.2, two ways to describe f as being continuous:

- $\forall$ open G in (Y, τ'), $f^{-1}(G)$ is open in (X, τ),
- $\forall$ closed F in (Y, τ'), $f^{-1}(F)$ is closed in (X, τ),

merely say that

- f^{-1} is an open map,
- f^{-1} is a closed map.

3.17

(ii) In (X, τ_{triv}), let $l \in X$ and $(x_n)_{n\geq 1}$ be any sequence. If N is 'any' neighbourhood of l then N must, in fact, be the whole of X! So

$$n \geq 1 \Rightarrow x_n \in N$$
$$\text{and therefore } x_n \to l.$$

(iv) In (X, τ_{disc}), suppose (x_n) converges to l.
Now $\{l\}$ is open (everything is open!) and is a neighbourhood of l, so
$\exists\, n_0 \in \mathbb{N}$ such that

$$\begin{aligned} n \geq n_0 &\Rightarrow x_n \in \{l\} \\ &\Rightarrow x_n = l. \end{aligned}$$

So (x_n) is eventually constant (at l).

(v) In (X, τ_{cc}), suppose (x_n) converges to l. Now $(X \setminus \{x_1, x_2, x_3, \ldots\}) \cup \{l\}$ has a countable complement, so it is τ_{cc}-open. Also, it contains l, so it is a neighbourhood of l. Hence

$$\exists n_0 \in \mathbb{N} \text{ such that } n \geq n_0 \Rightarrow x_n \in (X \setminus \{x_1, x_2, x_3, \ldots\}) \cup \{l\}$$
$$\Rightarrow x_n = l.$$

That is, (x_n) is eventually constant (at l).

3.18 Since $x \in G \in \tau$, G is a neighbourhood of x.
Now '$x_n \to x$' tells us $\exists n_0 \in \mathbb{N}$ such that

$$n \geq n_0 \Rightarrow x_n \in G$$

(that is, $x_n \in G \;\forall$ sufficiently large n).

3.19 Let N be any neighbourhood of l.
Since $x_n \to l, \exists n_0$ such that $n \geq n_0 \Rightarrow x_n \in N$.
In particular, $N \cap A \neq \emptyset$.
Now 2.14 tells us us that $l \in \overline{A}$.

3.20 Let N be any τ'-neighbourhood of $f(l)$.
Choose $H \in \tau'$ such that $f(l) \in H \subseteq N$.
Then $f^{-1}(H) \in \tau$ and $l \in f^{-1}(H)$, therefore $f^{-1}(H)$ is a τ-neighbourhood of l.
Since $x_n \to l, \exists n_0 \in \mathbb{N}$ such that

$$n \geq n_0 \Rightarrow x_n \in f^{-1}(H)$$
$$\Rightarrow f(x_n) \in H$$
$$\Rightarrow f(x_n) \in N.$$

So $f(x_n) \to f(l)$.

3.21 In $(\mathbb{R}, \tau_{cc})$ (remember what 3.17(v) told us!):
- $[0, 1]$ is not open, and yet if $x_n \to x \in [0, 1]$ then $x_n \in [0, 1]$ for all sufficiently large n.
- $2 \in \overline{[0, 1]}$ (indeed, $\overline{[0, 1]} = \mathbb{R}$!) and yet no sequence in $[0, 1]$ can have 2 as a limit.

- The identity map $id : (\mathbb{R}, \tau_{cc}) \to (\mathbb{R}, \tau_{usual})$ is not continuous: for example, because $(0, 1)$ is τ_{usual}-open but $id^{-1}((0, 1)) = (0, 1)$ is not τ_{cc}-open;
and yet $x_n \to x$ in $(\mathbb{R}, \tau_{cc}) \Rightarrow x_n$ eventually constant at x
$\Rightarrow x_n \to x$ in $(\mathbb{R}, \tau_{usual})$ also,
that is, $id(x_n) \to id(x)$ in $(\mathbb{R}, \tau_{usual})$.

3.25 'Only if' is 3.18.
Conversely, if G is *not* open then (see 2.6) $\exists x \in G$ s.t. G is not a neighbourhood of x.
Choose a countable local base $N_1 \supseteq N_2 \supseteq N_3 \supseteq \dots$ at x.
None of the N_i can be contained in G (otherwise, G would be a neighbourhood of x, *contradiction*)
so, for each $i \in \mathbb{N}$, we can select $x_i \in N_i \setminus G$.
The sequence $(x_i)_{i \geq 1}$ converges to x as limit ...
[CHECK: any neighbourhood J of x contains one of the N_i,
giving $J \supseteq N_i \supseteq N_{i+1} \supseteq N_{i+2} \supseteq \dots$
Therefore J includes $x_i, x_{i+1}, x_{i+2}, \dots$
therefore $x_i \to x$]
... and yet x_i never belongs to G:
that is, condition (1) fails.

3.26 'If' is 3.19.
Conversely, if $p \in \overline{A}$, choose a countable local base

$$N_1 \supseteq N_2 \supseteq N_3 \supseteq \dots$$

at p. For every i, the set N_i intersects A (see 2.14),
so we can choose $x_i \in A \cap N_i$.
This generates a sequence $(x_i)_{i \geq 1}$ of elements of A and, as in 3.25, $x_i \to p$.

3.27 'Only if' is 3.20.
Conversely, if f is NOT continuous, then (for example) there is a subset A of X such that $f(\overline{A}) \not\subseteq \overline{f(A)}$: see 3.2.
So choose $p \in \overline{A}$ such that $f(p) \notin \overline{f(A)}$.
By 3.26, there is a sequence $(a_n)_{n \geq 1}$ of points of A converging to p.

If $f(a_n)$ were to converge to $f(p)$, then 3.19 would assure us that $f(p) \in \overline{f(A)}$, *contradiction!*

$$\text{So} \quad f(a_n) \not\to f(p);$$
that is, (2) fails.

3.32 Let N_0 be any neighbourhood of x.
In the 'inverse inclusion' which orders $\mathcal{N}(x)$,

$$N \geq N_0 \Rightarrow N \subseteq N_0$$
$$\Rightarrow x_N \in N \subseteq N_0 :$$

that is (by Definition 3.31) $(x_N)_{N \in \mathcal{N}(x)} \to x$.

3.33 and 3.36 Suppose G is open.
If $x \in G$ and (net) $x_\gamma \to x$, then
G is a neighbourhood of x, so $\exists \gamma(G) \in$ [the underlying directed set D] such that

$$\gamma \geq \gamma(G) \Rightarrow x_\gamma \in G$$

(less formally, $x_\gamma \in G$ for all sufficiently large γ).
Suppose G is not open.
$\exists x \in G$ such that G is not a neighbourhood of x (2.6),
therefore no neighbourhood of x can be contained in G.
So $\forall N \in \mathcal{N}(x)$ we can choose $x_N \in N \setminus G$.
Then $(x_N)_{N \in \mathcal{N}(x)} \to x$ (3.32)
but x_N is *never* in G, so (1) fails.

3.34 and 3.37 Suppose there is a net $(x_\gamma)_{\gamma \in D}$ of elements of A which converges to p.
If N is any neighbourhood of p, then $x_\gamma \in N$ for all sufficiently large γ
and in particular $N \cap A \neq \emptyset$.
Via 2.14, $p \in \overline{A}$.

Suppose $p \in \overline{A}$.
Via 2.14, for each neighbourhood N of p we can choose $a_N \in N \cap A$.
This yields a net $(a_N)_{N \in \mathcal{N}(x)}$ of points of A,
and 3.32 tells us that $a_N \to p$.

3.35 and 3.38 Suppose f is continuous.

If we are given a net $(x_\gamma)_{\gamma \in D}$ converging to l and a τ'-neighbourhood U of $f(l)$,

we choose τ'-open G such that $f(l) \subseteq G \subseteq U$,

$$\begin{aligned} \text{observe} \quad & l \in f^{-1}(G) \in \tau \\ \text{so} \quad & f^{-1}(G) \text{ is a neighbourhood of } l, \\ \text{deduce} \quad & x_\gamma \in f^{-1}(G) \;\; \forall \gamma \geq \text{ some } \gamma_0, \\ \text{therefore} \quad & f(x_\gamma) \in G \subseteq U \quad \forall \gamma \geq \gamma_0, \end{aligned}$$

and conclude

$$\left(f(x_\gamma)\right)_{\gamma \in D} \to f(l).$$

Suppose f not continuous.

$$\begin{aligned} (3.2)\ \exists\, A \subseteq X \text{ such that } & f(\overline{A}) \not\subseteq \overline{f(A)}, \\ \exists\, p \in \overline{A} \text{ such that } & f(p) \notin \overline{f(A)}, \\ (3.34)\ \exists \text{ net } \left(a_\gamma\right)_{\gamma \in D} \text{ in } A & \text{ converging to } p. \end{aligned}$$

But if $f(a_\gamma) \to f(p)$, then $f(p)$ would be in the closure of $f(A)$ (3.37) – *contradiction!*

So $f(a_\gamma) \not\to f(p)$ and (2) fails.

3.42

(i) Suppose $p \in \overline{A}$.

Put $\Lambda = \{F \subseteq X : \exists N \in \mathcal{N}(p) \text{ such that } F \supseteq A \cap N\}$.

Then Λ is a filter on X

and $\mathcal{N}(p) \subseteq \Lambda$

and $A \in \Lambda$.

(ii) Suppose there is a filter Λ as described.

Then for every neighbourhood N of p, $N \cap A \in \Lambda$, therefore $N \cap A \neq \emptyset$.

Therefore $p \in \overline{A}$.

3.43 Notice first that a space is compact if and only if, whenever every finite sub-family of a family of closed subsets has non-empty intersection, then the entire family also has non-empty intersection. (Proof: take complements across the definition of compactness, and use De Morgan's laws.)

(i) Suppose (X, τ) is compact.
Given a filter Λ, notice that $\{\overline{F} : F \in \Lambda\}$ is a family of closed sets, the intersections of whose finite subfamilies are all non-empty.
Therefore $\exists p \in \bigcap\{\overline{F} : F \in \Lambda\}$.
Put $\Lambda_1 = \{$all supersets of sets of the form $F \cap N : F \in \Lambda, N \in \mathcal{N}(p)\}$.
Then Λ_1 is a filter on X, $\mathcal{N}(p) \subseteq \Lambda_1, \Lambda \subseteq \Lambda_1$ as required.

(ii) Suppose (X, τ) is not compact.
Then there is a family Γ of closed subsets whose finite subfamilies all have non-empty intersections, and yet the entire family has empty intersection.
Then $\Lambda =$ the collection of all supersets of finite intersections from Γ is a filter on X.
If, for some filter Λ', we had both $\Lambda \subseteq \Lambda'$ and $\mathcal{N}(p) \subseteq \Lambda'$,
then for all $N \in \mathcal{N}(p)$ and for all $F \in \Gamma$, $N \cap F \neq \emptyset$.
Therefore $p \in \overline{F} = F$ for all $F \in \Gamma$ – *contradiction.*

3.44

(i) $\Lambda(x)$ converges to p if and only if $\mathcal{N}(p) \subseteq \Lambda(x)$
if and only if every neighbourhood of p contains a tail of the net
if and only if the net x converges to p.

(ii) (a) If Λ converges to p
then every neighbourhood N of p contains some $A \in \Lambda$,
therefore $N \supseteq B$ for all $B \geq A$ (that is, for all $B \subseteq A$),
therefore $x_B \in N$ for all $B \geq A$,
that is, the net x converges to p.

(b) If Λ does not converge to p
then some neighbourhood N of p does not belong to Λ,
therefore N does not contain any $A \in \Lambda$,
so for each $A \in \Lambda$ we can select $x_A \in A \setminus N$
and this creates an appropriate net, failing to converge to p.

3.46

(i) Suppose f is continuous.
Let the filterbase Φ converge to p
(that is, each neighbourhood of p contains some $F \in \Phi$).

If N is a neighbourhood of $f(p)$ in Y
then $f^{-1}(N)$ is a neighbourhood of p in X
so $f^{-1}(N)$ contains some $F \in \Phi$,
that is, $N \supseteq f(F)$;
therefore $f(\Phi)$ converges to $f(p)$.

(ii) Suppose f is not continuous.
Then $\exists$ closed $K \subseteq Y$ such that $f^{-1}(K)$ is not closed in X,
so $\exists p \in \overline{f^{-1}(K)}$ such that $p \notin f^{-1}(K)$.
Then $\Phi = \{N \cap f^{-1}(K) : N \in \mathcal{N}(p)\}$ is a filterbase that converges to p.
Yet $f(\Phi)$ does not converge to $f(p)$, since $Y \setminus K$ is a neighbourhood of $f(p)$ containing no $f(F)$ for $F \in \Phi$.

Invariants

We explained in 3.8 the meaning of the term 'topological property'/'homeomorphic invariant' (let us call it 'invariant' for short). First-countability is an example of an invariant, as we can confirm by either of two methods:

(a) by meticulously checking that if

$$h : (X, \tau) \rightarrow (Y, \tau')$$

is a homeomorphism and (X, τ) is first-countable, then also (Y, τ') is first-countable; or

(b) by observing that the definition (3.24, 3.22) contains nothing but *open sets* plus *basic set theory* plus *ideas that can be fully described in terms of these alone.*

We now identify and examine a handful of the most useful and important invariants. Recurrent questions to ask of each will include:

(a) when do subspaces possess the invariant?
(b) how do continuous maps relate to it?
(c) does it behave more simply in metric spaces?

Sequential compactness

4.1 Definition A space (X, τ) is *sequentially compact* if, in X, every sequence has a convergent subsequence (with limit in X, of course).

4.2 Examples Trivial spaces are sequentially compact; $(\mathbb{R}, \tau_{cc})$ is not.

4.3 Proposition Sequential compactness is closed-hereditary.

We recall that a subset of a metric space is called *bounded* if the set of all point-to-point distances between elements of that subset is bounded (above).

4.4 Proposition Let (M, d) be a metric space and A a subset of M, and suppose that the subspace $(A, (\tau_d)_A)$ is sequentially compact. Then A is both closed and bounded in M.

4.5 Note The converse of 4.4 is false in general: consider the 'flattened' metric $\hat{d}$ defined on $\mathbb{R}$ by

$$\hat{d}(x, y) = \min\{1, |x - y|\}$$

and note that $\mathbb{R}$ itself is closed and bounded under $\hat{d}$, but not sequentially compact (in the topology that it induces, which is, in fact, τ_{usual} again). However:

4.6 Proposition Give $\mathbb{R}$ its usual metric. Then a subspace of $\mathbb{R}$ is sequentially compact if and only if it is closed and bounded.

4.7 Notes

(1) It is efficient to say 'A is a sequentially compact subset of (X, τ)' rather than the more correct but clumsy 'A is a subset of (X, τ) such that the subspace (A, τ_A) is sequentially compact.' This is standard practice for invariants.

(2) Sequential compactness is not open-hereditary (and therefore not hereditary): think of intervals on the real line and invoke 4.6.

4.8 Proposition Sequential compactness is *preserved by continuous surjections:* that is, given that $f : (X, \tau) \to (Y, \tau')$ is continuous and onto, and that (X, τ) is sequentially compact, it follows that (Y, τ') is sequentially compact.

Compactness

4.9 Definitions An *open cover* of a subset A in a space (X, τ) is a family of τ-open sets whose union contains A. A *finite subcover* (of such a cover) is a selection of finitely many of these sets whose union still contains all of A. The special case in which $A = X$ is particularly important: in that scenario, we speak of an open cover *of the space.*

4.10 Examples

(i) The family of open intervals $\{(-n, n) : n \geq 1\}$ is an open cover of the real line (under its usual topology) possessing no finite subcover.

(ii) For each $t \in (0, 1)$ let G_t denote $(t, 1]$, and let $G_0 = [0, 1/1000)$. Then all these sets make up an open cover for the subspace $[0, 1]$ of the real line. There is a finite subcover.

4.11 Definition A space is compact when *every* one of its open covers has a finite subcover.

4.12 Examples The real line is not compact. Any cofinite space (X, τ_{cf}) is compact.

In accordance with 4.7, the phrase 'A is a compact subset of a space (X, τ)' means that the subspace (A, τ_A) is compact: that is, every τ_A-open cover of A has a finite subcover (for A). It is a nuisance to have to look at τ_A rather than τ here, especially if (as often happens) we are dealing with more than one subset at a time. Luckily, we do not have to:

4.13 Lemma The set A is a compact subset of the space (X, τ) if and only if every τ-open cover of A has a finite subcover.

4.14 Proposition Compactness is closed-hereditary.

4.15 Proposition Compactness is preserved by continuous surjections.

The precise relationship between sequential compactness and compactness is tricky to sort out. The good news is: for metrisable spaces, they are entirely equivalent.

4.16 Lemma Let S be any infinite subset of a compact topological space X. There is some $a \in X$ such that every neighbourhood of a contains infinitely many points of S.

4.17 The 'tapioca' or 'frogspawn' lemma Given a sequentially compact metric space (M, d) and a positive real number δ, there is a finite family of open balls each of radius δ that covers M (Fig. 4.1).

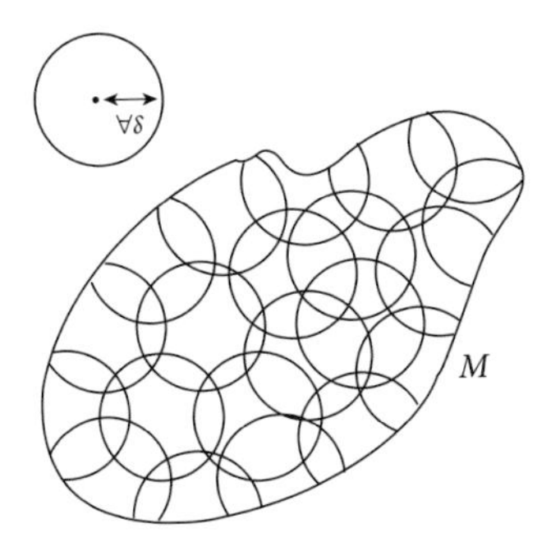

Fig. 4.1 The frogspawn lemma (see 4.17).

4.18 Theorem A metrisable space is compact if and only if it is sequentially compact.

4.19 Corollary

(i) Compact subsets of a metric space must be closed and bounded.

(ii) The converse of (i) is false.

(iii) In $\mathbb{R}$ with the usual metric, the compact subsets are precisely the closed and bounded ones.

(iv) The Cantor set is compact.

4.20 Note You probably know that, for each positive integer n, 4.19(iii) is true also for $\mathbb{R}^n$ under its usual metric. If not, we'll prove it in the next chapter.

Local compactness

4.21 Definition We call a space (X, τ) *locally compact* if each point in X has a compact neighbourhood.

4.22 Examples

(i) The real line and its 'finite powers' are locally compact (though not compact).

(ii) Any compact space is locally compact.

(iii) The set $\mathbb{Q}$ of rationals, as a subspace of $(\mathbb{R}, \tau_{\text{usual}})$, is not locally compact.

4.23 Proposition Local compactness is closed-hereditary.

4.24 Proposition Local compactness is preserved by continuous open surjections.

4.25 Note Local compactness is not preserved by all continuous surjections: consider the identity map from $(\mathbb{Q}, \tau_{\text{disc}})$ to $(\mathbb{Q}, \tau_{\text{usual}})$.

Connectedness

4.26 Definition A *partition* of (X, τ) is a pair of non-empty disjoint τ-open sets whose union is X. (Note that if P, Q are disjoint and their union is X, then each is the complement of the other: so they are both open if and only if they are both closed!) A *connected* space is one which has no partition. Once again, a connected subset A of (X, τ) is one for which (A, τ_A) is connected. Spaces and subsets that are not connected can be called *disconnected* or *non-connected*.

4.27 Lemma The space (X, τ) is connected if and only if X and $\emptyset$ are the only subsets of X that are both open and closed.

4.28 Examples

(i) Trivial spaces are connected, discrete spaces (except on singletons) are not.

(ii) $(\mathbb{Q}, \tau_{\text{usual}})$ is not connected: for instance, because its proper subset $\mathbb{Q} \cap [\sqrt{2}, \sqrt{3}] = \mathbb{Q} \cap (\sqrt{2}, \sqrt{3})$ is both open and closed in the subspace $\mathbb{Q}$.

(iii) A cofinite space on an infinite set is connected, and so is a cocountable space on an uncountable set.

4.29 Lemma The non-null subset A of (X, τ) is non-connected if and only if:

$$\exists\tau\text{-open } G,\ H \text{ such that } A \subseteq G \cup H,\ A \cap G \neq \emptyset,\ A \cap H \neq \emptyset,\ A \cap G \cap H = \emptyset$$

(Fig. 4.2).

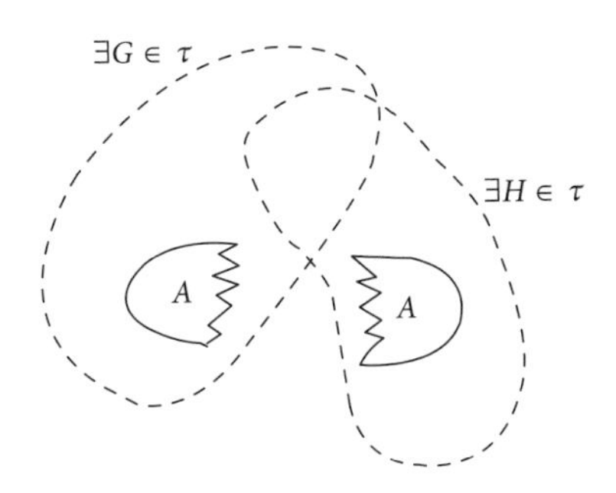

Fig. 4.2 A is a non-connected subset (see 4.29).

4.30 Note That may look unpleasant, but it has the valuable feature of identifying subset-connectedness in terms of the background topology rather than of the subspace topology: compare 4.13. Again we can replace 'open' by 'closed' here.

4.31 Note By an *interval* in $\mathbb{R}$ we mean any subset $I \subseteq \mathbb{R}$ which has the property that $a < b < c,\ a \in I,\ c \in I$ together imply $b \in I$. It is routine to verify that this definition agrees with the usual listing of all possible types (open, closed, bounded, ...) of real interval. It will surprise no one that these are exactly the connected subsets of the real line:

4.32 Lemma In $\mathbb{R}$, if $[a, b] = F_1 \cup F_2$, where F_1, F_2 are closed and $a \in F_1$ and $b \in F_2$, then $F_1 \cap F_2 \neq \emptyset$.

4.33 Proposition A subset of the real line (with the usual topology) is connected if and only if it is an interval.

4.34 Example No subset of the Cantor set – with the (inevitable) exception of the one-element subsets – is connected.

4.35 Proposition Connectedness is preserved by continuous surjections.

This insight allows us to revisit two elementary results of real analysis, and understand them as topological in essence:

4.36 Corollary 1: the 'intermediate value theorem' If $f : [a, b] \to \mathbb{R}$ is continuous and y lies between $f(a)$ and $f(b)$, then y is a value of f also.

4.37 Corollary 2: the 'fixpoint theorem for $[0, 1]$' Any continuous map from $[0, 1]$ to $[0, 1]$ has a fixed point.

4.38 Proposition The closure of a connected set is connected. More generally, if A is a connected subset of (X, τ) and $A \subseteq B \subseteq \overline{A}$, then B is connected.

Separability

4.39 Definitions A subset D of a space (X, τ) is called *dense* if $\overline{D} = X$. A space (X, τ) is called *separable* if it possesses a countable dense subset.

4.40 Examples $(\mathbb{R}, \tau_{\text{usual}})$ and $(\mathbb{R}, \tau_{\text{cf}})$ are separable, but $(\mathbb{R}, \tau_{\text{cc}})$ is not.

4.41 Proposition Separability is open-hereditary.

4.42 Note Separability is not closed-hereditary (and so, not hereditary). Under an included-point topology, say, ι_2, the set $\mathbb{R}$ becomes separable in a rather extreme fashion. Yet its ι_2-closed subset $\mathbb{R} \setminus \{2\}$ is not separable.

4.43 Proposition Separability is preserved by continuous surjections.

We shall have more to say about separability in the next chapter.

Exercises Essential Exercises 30–43 are based on the material in this chapter. It is particularly recommended that you should try numbers 33, 34, 35, 37, 40, 42 and 43.

Expansion of Chapter 4

4.2 Any sequence in a trivial space is convergent already (3.17(ii)).
In $(\mathbb{R}, \tau_{\text{cc}})$ take, for example, the sequence $\left(\frac{1}{n}\right)_{n \geq 1}$.
It is not eventually constant, and *none* of its subsequences is eventually constant either.
So (3.17(v)) none of its subsequences converges.

4.3 Let (X, τ) be sequentially compact, $\emptyset \neq A \subseteq X$, A being closed in (X, τ).
Consider any sequence of points $(a_n)_{n \geq 1}$ in A.
Now (a_n) is a sequence in (sequentially compact) X,
so it has a convergent subsequence in X,
that is, some subsequence $(a_{n_k}) \to$ some ℓ in X.
But A is closed, so $\ell \in \overline{A} = A$ (see 3.19, for example),
so (a_{n_k}) converges to ℓ within (A, τ_A).
Hence (A, τ_A) is sequentially compact also.

4.4 If *A were not closed:*

we could (see 3.26) find a sequence (a_n) in A and a point $p \in \overline{A} \setminus A$ such that $a_n \to p$.

Since A is sequentially compact, some subsequence (a_{n_k}) must converge to a limit in A: say, $a_{n_k} \to q \in A$.

But (standard results on sequences in metric spaces!) since $a_n \to p$, also (the subsequence) $a_{n_k} \to p$

and limits (in metric spaces) are unique, so

$$p = q \text{ where } p \notin A \text{ and } q \in A \ldots$$

contradiction!

If A were not bounded:

choose a_0 anywhere in A. For each $n \in \mathbb{N}$ there must exist $a_n \in A$ such that $d(a_0, a_n) > n$

(for otherwise the triangle inequality implies that no two points of A can be more than $2n$ apart, *contradiction* to 'not bounded').

The sequence (a_n) and all its subsequences will then be unbounded, and cannot converge.

So this *contradicts* sequential compactness.

4.6 'Sequential compactness $\Rightarrow$ closed and bounded' is just 4.4.

'Closed and bounded in $\mathbb{R} \Rightarrow$ sequentially compact' is the Bolzano–Weierstrass result of elementary real analysis.

4.8 Let $(y_n)_{n\geq 1}$ be any sequence in (Y, τ').

Since f is onto, for each n choose $x_n \in X$ such that $f(x_n) = y_n$.

The sequence $(x_n)_{n\geq 1}$ in X has a convergent subsequence (x_{n_k}):

$$x_{n_k} \to \ell \quad \text{in } X.$$

Since f is continuous, $f(x_{n_k}) \to f(\ell)$ in Y, that is, $y_{n_k} \to f(\ell)$.

Thus (y_n) has a convergent subsequence,

and (Y, τ') is sequentially compact.

4.12

- The sequence of intervals $\{(-n, n) : n \in \mathbb{N}\}$ is an open cover of $(\mathbb{R}, \tau_{\text{usual}})$ and it has no finite subcover.

- Given $\{G_\alpha : \alpha \in I\}$ an open cover of (X, τ_{cf}),
  take any α_0 so that $G_{\alpha_0} \neq \emptyset$.
  Since G_{α_0} is τ_{cf}-open, $X \setminus G_{\alpha_0}$ is a *finite* set, say,

$$X \setminus G_{\alpha_0} = \{x_1, x_2, \ldots, x_n\}.$$

For each $i = 1, 2, \ldots, n$ choose $\alpha_i \in I$ so that $x_i \in G_{\alpha_i}$.
Now $\left\{G_{\alpha_0}, G_{\alpha_1}, G_{\alpha_2}, \ldots, G_{\alpha_n}\right\}$ covers all of X.
Hence (X, τ_{cf}) is compact.

4.13

(i) Suppose A is a compact subset of (X, τ), that is, (A, τ_A) is compact.
Let $\{G_\alpha : \alpha \in I\}$ be any τ-open cover of A.
Then $\{A \cap G_\alpha : \alpha \in I\}$ is a τ_A-open cover of A, and therefore has a finite subcover $\{A \cap G_{\alpha_1}, A \cap G_{\alpha_2}, \ldots, A \cap G_{\alpha_n}\}$.
So $A \subseteq \bigcup_1^n G_{\alpha_i,}$ that is, $\{G_{\alpha_1}, \ldots, G_{\alpha_n}\}$ is a finite subcover of the given cover.

(ii) Suppose every τ-open cover of A has a finite subcover.
Let $\{H_\alpha : \alpha \in I\}$ be any τ_A-open cover of A.
Each H_α takes the form $A \cap G_\alpha$, some $G_\alpha \in \tau$.
But then $\{G_\alpha : \alpha \in I\}$ is a τ-open cover of A, so it must have a finite subcover $\{G_{\alpha_1}, G_{\alpha_2}, \ldots G_{\alpha_n}\}$.
Then $A = A \cap \bigcup_1^n G_{\alpha_i} = \bigcup_1^n (A \cap G_{\alpha_i}) = \bigcup_1^n H_{\alpha_i}$,
that is, $\left\{H_{\alpha_1}, \ldots, H_{\alpha_n}\right\}$ is a finite subcover of A.
Hence (A, τ_A) is compact.

4.14 Let (X, τ) be compact, $\emptyset \neq A \subseteq X$, A being τ-closed.
For any τ-open cover $\{G_\alpha : \alpha \in I\}$ of A,
$\{X \setminus A, G_\alpha : \alpha \in I\}$ is a τ-open cover of all of X
so there is a finite subcover $\{X \setminus A, G_{\alpha_1}, G_{\alpha_2}, \ldots, G_{\alpha_n}\}$.
Then $A \subseteq \bigcup_1^n G_{\alpha_i}$
so $\{G_{\alpha_1}, \ldots, G_{\alpha_n}\}$ is a finite subcover of A.
By 4.13, A is compact.

4.15 Let $f : (X, \tau) \to (Y, \tau')$ be continuous and onto, where (X, τ) is compact.
Given $\{H_\alpha : \alpha \in I\}$ a τ'-open cover of Y,

$$X = f^{-1}(Y) = f^{-1}\left(\bigcup_\alpha H_\alpha\right) = \bigcup_\alpha f^{-1}(H_\alpha),$$

that is, $\{f^{-1}(H_\alpha) : \alpha \in I\}$ is an *open* cover of X.
There is a finite subcover $\{f^{-1}(H_{\alpha_i}) : 1 \le i \le n\}$.
Now

$$Y = f(X) = f\left(\bigcup_1^n f^{-1}(H_{\alpha_i})\right)$$

$$= \bigcup_1^n ff^{-1}(H_{\alpha_i})$$

$$\subseteq \bigcup_1^n H_{\alpha_i}$$

('equals', actually, since f is onto!),
that is, $\{H_{\alpha_i} : 1 \le i \le n\}$ is a finite subcover.
So Y is compact.

4.16 *If not* then, for each $a \in X$, we find a neighbourhood N_a of a such that $N_a \cap S$ is finite.
Inside N_a, find an open set G_a (such that $a \in G_a \subseteq N_a$)
and then $G_a \cap S$ is also finite.
Now $\{G_a : a \in X\}$ is an open cover of compact X
so $\exists$ some finite subcover:

$$X = \bigcup_1^n G_{a_i}.$$

Then $S = S \cap X = S \cap \bigcup_1^n G_{a_i} = \bigcup_1^n (S \cap G_{a_i})$
is a finite union of finite sets,
therefore S is finite – *contradiction!*

4.17 *If not*, then no finite number of open balls $B(p, \delta)$ of radius δ can possibly cover all of M.
Choose x_1 anywhere in M.
Choose $x_2 \notin B(x_1, \delta)$.
Choose $x_3 \notin B(x_1, \delta) \cup B(x_2, \delta)$
. . . and so on.

When we have chosen $x_1, x_2, \ldots, x_n$ it will still be the case that $\bigcup_1^n B(x_i, \delta)$ is *not all* of M,
so we can choose $x_{n+1} \notin \bigcup_1^n B(x_i, \delta)$.

This generates an (infinite) sequence $(x_i)_{i\geq 1}$, every two terms of which are at least δ apart.
So *no* subsequence of (x_i) can be Cauchy,
so *no* subsequence of (x_i) can be convergent: *contradiction!*

4.18

(i) Let (M, d) be compact (metric).
Let $(x_n)_{n\geq 1}$ be any sequence in M.

- If some element of M occurs infinitely often in the sequence, its occurrences form a subsequence that is constant, and therefore certainly convergent.

- If not, then $S = \{x_n : n \in \mathbb{N}\}$ is an infinite set.
By 4.16, $\exists\, a$ such that every $B\left(a, \frac{1}{k}\right)$ contains infinitely many x_n's.

Choose n_1 so that $B(a, 1)$ includes x_{n_1}.
Choose $n_2 > n_1$ so that $B(a, \frac{1}{2})$ includes x_{n_2}.
Choose $n_3 > n_2$ so that $B\left(a, \frac{1}{3}\right)$ includes x_{n_3}
... and so on.

Clearly this process runs for ever, and generates a subsequence $\left(x_{n_k}\right)_{k\geq 1}$ for which

$$d(x_{n_k}, a) < \frac{1}{k} \to 0 \quad (\text{as } k \to \infty),$$

that is, $x_{n_k} \to a$.
Hence (M, d) is sequentially compact.

(ii) Suppose (M, d) is sequentially compact.
If possible, let $\{G_\alpha : \alpha \in I\}$ be a given open cover of M, having *no* finite subcover.
For each $n \in \mathbb{N}$ use 4.17 to cover M by open balls – finitely many! – of radius $\frac{1}{n}$.
At least one of them ... call it $B(x_n, \frac{1}{n})$ ... can't be covered by any finite number of the G_α's
(otherwise, so would M be covered).
The sequence $(x_n)_{n\geq 1}$ thus created has a convergent subsequence

$$x_{n_k} \to \ell.$$

Now $\ell \in$ some G_{α_0}, and $\exists \varepsilon > 0$ so that

$$B(\ell, \varepsilon) \subseteq G_{\alpha_0}.$$

But as $k \to \infty$, both $d(x_{n_k}, \ell)$ and $\frac{1}{n_k}$ will $\to 0$, which, for sufficiently big

values of k, will make $B\left(x_{n_k}, \frac{1}{n_k}\right)$ lie entirely inside $B(\ell, \varepsilon)$

and therefore inside G_{α_0}, in contradiction to how the $B(x_n, \frac{1}{n})$'s were chosen.
So $\{G_\alpha : \alpha \in I\}$ has to have a finite subcover,
and (M, d) is compact.

4.22

(i) $\forall x \in \mathbb{R}$, $[x-1, x+1]$ is a compact neighbourhood of x in $(\mathbb{R}, \tau_{\text{usual}})$.
$\forall x \in \mathbb{R}^n$, $\overline{B}(x,1) = \{y \in \mathbb{R}^n : d(x,y) \leq 1\}$ is a closed and bounded (therefore compact! see 4.19(iii)) neighbourhood of x.

(ii) If X itself is compact then, for each $x \in X$, X is a compact neighbourhood of x; so X is locally compact.

(iii) Suppose $\mathbb{Q}$ were locally compact. Then 0 has a compact neighbourhood N and we can choose $\varepsilon > 0$ so that $[-\varepsilon, \varepsilon] \cap \mathbb{Q} \subseteq N$.
Since compactness is closed-hereditary and $[-\varepsilon, \varepsilon] \cap \mathbb{Q}$ is closed in $\mathbb{Q}$, and therefore also closed in N, $[-\varepsilon, \varepsilon] \cap \mathbb{Q}$ will be compact also.
This implies that it is compact also as a subset of $\mathbb{R}$, and closed in $\mathbb{R}$.
But it is not! Look:

$$\overline{[-\varepsilon, \varepsilon] \cap \mathbb{Q}}^{\mathbb{R}} = [-\varepsilon, \varepsilon].$$

4.23 Let (X, τ) be locally compact, $\emptyset \neq A \subseteq X$, A being τ-closed, and $x \in A$.
Now x has a compact τ-neighbourhood N.
Then $A \cap N$ is a τ_A-neighbourhood of x (2.17).
Also, $A \cap N$ is τ_N-closed (2.16)
so it inherits compactness from N (4.14).
So $A \cap N$ is compact as a subset of A and is a τ_A-neighbourhood of x:
therefore (A, τ_A) is locally compact.

4.24 Let (X, τ) be locally compact, and $f : (X, \tau) \to (Y, \tau')$ continuous, open, onto.

Given $y \in Y$,
choose $x \in X$ such that $f(x) = y$,
choose compact neighbourhood N of x,
choose τ-open G such that $x \in G \subseteq N$.

Now $f(x) \in f(G) \subseteq f(N)$, where
$f(x) = y$, $f(G)$ is τ'-open, $f(N)$ is compact (4.15),
that is, $f(N)$ is a compact neighbourhood of y,
so Y is locally compact.

4.27

(i) If X is *not* connected, it has a partition

$$X = P \cup Q, P \cap Q = \emptyset, P \in \tau, Q \in \tau, P \neq \emptyset, Q \neq \emptyset$$

and then P is open and closed and is neither empty nor the whole of X.

(ii) If $\exists A$ neither empty nor X that is both open and closed,
then A and $X \setminus A$ form a partition of X
so X is not connected.

4.29

(i) If A is not connected then
(A, τ_A) has a partition:
$A = B \cup C, B \cap C = \emptyset, B$ and C are non-empty and τ_A-open.
Now B, C take the form $A \cap G, A \cap H$, where $G, H \in \tau$.
These G, H are as required!

(ii) If G, H exist as described then

$$A = A \cap (G \cup H) = (A \cap G) \cup (A \cap H),$$

where $(A \cap G) \cap (A \cap H) = \emptyset, A \cap G$ and $A \cap H$ are non-empty and τ_A-open,
that is, $A \cap G, A \cap H$ make a partition of (A, τ_A),
so A is not connected.

4.32 F_1 is non-empty and bounded above (by b)
so it has a supremum $\lambda \leq b$.

If $\lambda = b$, then $\forall n \in \mathbb{N}\ \exists x_n \in F_1$ such that $b - \frac{1}{n} < x_n \leq b$,

$$\begin{aligned} &\text{therefore} \quad x_n \to b, \\ &\text{therefore} \quad b \in \overline{F_1} = F_1 \text{ and also } b \in F_2, \\ &\text{so } F_1 \cap F_2 \neq \emptyset. \end{aligned}$$

If $\lambda < b$, then *as above* $\lambda \in F_1$,

$$\begin{aligned} &\text{but also } (\lambda, b] \subseteq F_2, \\ &\text{therefore } \lambda \in \overline{F_2} = F_2, \\ &\text{so again } F_1 \cap F_2 \neq \emptyset. \end{aligned}$$

4.33

(i) If $A \subseteq \mathbb{R}$ is not an interval,
then $\exists a_1, b, a_2 \in \mathbb{R}$ such that
$a_1 < b < a_2, a_1 \in A, a_2 \in A, b \notin A$.
Put $G = (-\infty, b), H = (b, \infty)$,
and 4.29 shows that A is not connected.

(ii) If $A \subseteq \mathbb{R}$ is not connected, then (via 4.29, 'closed' version) $\exists \tau$-closed K_1, K_2 such that

$$A \subseteq K_1 \cup K_2, A \cap K_1 \neq \emptyset \neq A \cap K_2 \text{ and } A \cap K_1 \cap K_2 = \emptyset.$$

Pick $a \in A \cap K_1, b \in A \cap K_2$
(without loss of generality, $a < b$)
and put $F_1 = [a, b] \cap K_1, F_2 = [a, b] \cap K_2$.

If A were an interval
we would have
$[a, b] \subseteq A$ and $[a, b] \subseteq K_1 \cup K_2$, that is, $[a, b] = F_1 \cup F_2$,
so 4.32 would give $\emptyset \neq F_1 \cap F_2 \subseteq A \cap K_1 \cap K_2 = \emptyset$: *contradiction!*
So A *cannot* be an interval.

4.34 This follows from 4.33 since, as we pointed out in 3.10, the Cantor set does not contain any non-degenerate interval.

4.35 Let $f : (X, \tau) \to (Y, \tau')$ be continuous and onto, where X is connected.
If Y is not connected, partition it:
$Y = G \cup H, G \cap H = \emptyset, G$ and $H \in \tau', G$ and H non-empty.

Then $f^{-1}(G), f^{-1}(H)$ are open, disjoint, non-empty, and their union is X, that is, X is partitioned.
Contradiction!
So Y has to be connected.

4.36 Given $f: [a, b] \to \mathbb{R}$ continuous
$[a, b]$ is connected (4.33),
therefore $f([a, b])$ is connected (4.35),
therefore $f([a, b])$ is an interval (4.33),
that is, any number lying between two elements of $f([a, b])$ is an element of $f([a, b])$.

4.37 Let $f: [0, 1] \to [0, 1]$ be continuous.
Define $g: [0, 1] \to \mathbb{R}$ by $g(x) = x - f(x)$.
Then g is also a continuous real function, and

$$g(0) = -f(0) \leq 0 \leq 1 - f(1) = g(1).$$

Therefore 4.36 says 0 is a value of g,

$$\text{that is, } \exists t \in [0, 1] \text{ such that } 0 = g(t) = t - f(t),$$
$$\text{that is, } f(t) = t.$$

4.38 Given A connected in (X, τ) and $A \subseteq B \subseteq \overline{A}$,
suppose B is *not* connected.
By 4.29, $\exists$ open G, H such that

$$B \subseteq G \cup H, B \cap G \cap H = \emptyset, B \cap G \neq \emptyset \text{ and } B \cap H \neq \emptyset. \quad (1)$$

Now choose $b \in B \cap G$ and observe that
$b \in \overline{A}$ and G is an open neighbourhood of b.
Therefore by 2.14, $A \cap G \neq \emptyset$.
Similarly, $A \cap H \neq \emptyset$.
So (1) gives $A \subseteq G \cup H, A \cap G \cap H = \emptyset$ as well as $A \cap G \neq \emptyset \neq A \cap H$,
that is, A is not connected . . . *contradiction!*

4.40

- In $(\mathbb{R}, \tau_{\text{usual}})$, $\mathbb{Q}$ is countable and dense.

- In $(\mathbb{R}, \tau_{cf})$, any countably infinite set . . . for example, $\mathbb{Z}$ . . . is countable and dense, because its closure must be τ_{cf}-closed and cannot be finite.

- In $(\mathbb{R}, \tau_{cc})$, any countable set is closed, so equals its own closure, so cannot be dense.

4.41 Let (X, τ) be separable, $\emptyset \neq G \subseteq X$ be τ-open.
Choose countable dense D in X.
Certainly $D \cap G$ is a countable subset of G, but is it dense?
For any $x \in G$ and any τ-neighbourhood N of x,
$G \cap N$ is also a τ-neighbourhood of $x \in \overline{D}$
so $(G \cap N) \cap D \neq \emptyset$,
that is, $N \cap (D \cap G) \neq \emptyset$.
That shows $x \in \overline{D \cap G}^{\tau}$
and therefore $G \subseteq \overline{D \cap G}^{\tau}$. Then

$$\overline{D \cap G}^{\tau_G} = G \cap \overline{D \cap G}^{\tau} \quad (2.18)$$
$$= G,$$

that is, in (G, τ_G), $D \cap G$ is dense.

4.42 In $(\mathbb{R}, \iota_2)$, $\overline{\{2\}}$ must be the whole of $\mathbb{R}$ since no other closed set includes 2. Yet the subspace $\mathbb{R} \setminus \{2\}$ is discrete: for each $x \neq 2$ in $\mathbb{R}$, $\{x, 2\}$ is ι_2-open, so $\{x\} = (\mathbb{R} \setminus \{2\}) \cap \{x, 2\}$ is open in the subspace. Now that we know all singletons are open, *every* subset (the union of all its own singletons) must also be open: that is, the subspace topology on this set is the discrete topology.

4.43 Let $f : (X, \tau) \to (Y, \tau')$ be continuous and onto, where X is separable.
Choose countable dense $D \subseteq X$, that is, $\overline{D} = X$.
Certainly $f(D) \subseteq Y$ is countable. By 3.2:

$$f(\overline{D}) \subseteq \overline{f(D)},$$
that is, $$Y = f(X) \subseteq \overline{f(D)} \subseteq Y,$$
so $$Y = \overline{f(D)}$$ and $f(D)$ is dense in Y.

Base and product

Base

Not every open set in $(\mathbb{R}, \tau_{\text{usual}})$ is an open interval, but every open set is a *union* of open intervals. Because of this, in many 'calculations' with open sets it is *enough* to work with open intervals: they are simple, and they are the basic building blocks from which all open sets can be constructed. Identical remarks apply to open discs in $(\mathbb{R}^2, \tau_{\text{usual}})$, open balls in $(\mathbb{R}^3, \tau_{\text{usual}}) \ldots$, and we now generalise the notion.

5.1 Definitions

(a) A *base* for a topological space (X, τ) (or for its topology τ) is a collection $\mathcal{B}$ of subsets of X such that

  (i) all the sets in $\mathcal{B}$ are τ-open, and

  (ii) every non-empty τ-open set can be expressed as a union of members of $\mathcal{B}$.

(b) A *subbase* for a topological space (X, τ) (or for its topology τ) is a collection $\mathcal{S}$ of subsets of X such that

  (i) all the sets in $\mathcal{S}$ are τ-open, and

  (ii) all the finite intersections of members of $\mathcal{S}$ form a base for (X, τ).

5.2 Examples The open intervals form a base for $(\mathbb{R}, \tau_{\text{usual}})$. In any metric space, the open balls form a base for the induced topology. In a discrete space, the collection of singletons is a base. For harder examples, the following lemma is useful:

5.3 A 'recognition lemma' If $\mathcal{B}$ is a collection of open sets of the space (X, τ), then $\mathcal{B}$ is a base for τ if and only if:

$$x \in G \in \tau \Rightarrow \exists B \in \mathcal{B} \text{ such that } x \in B \subseteq G \tag{1}$$

(Fig. 5.1).

5.4 Example The Cantor set has a base consisting of sets that are both open and closed.

Here are a couple of illustrations of how to use a base in a topological argument.

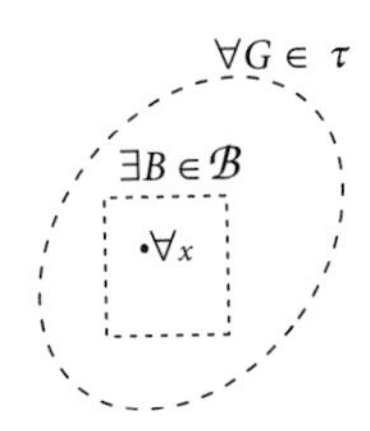

Fig. 5.1 How to recognise a base (see 5.3).

5.5 Lemma Suppose that $f : (X, \tau) \to (Y, \tau')$, $\mathcal{B}$ is a base for (Y, τ') and $f^{-1}(B)$ is τ-open for every $B \in \mathcal{B}$. Then f is continuous.

5.6 Lemma In a space (X, τ), let p be a point, A be a subset and $\mathcal{B}$ be a base. Then $p \in \overline{A}$ if and only if each member of $\mathcal{B}$ that includes p intersects A.

5.7 Lemma In a space (X, τ), let (A, τ_A) be a subspace and $\mathcal{B}$ be a base. Then

$$\{A \cap B : B \in \mathcal{B}\}$$

is a base for τ_A.

Complete separability

5.8 Definition We call (X, τ) *completely separable* if the topology has a countable base.

5.9 Examples The (Euclidean) real line is completely separable, but an included-point topology on $\mathbb{R}$ is not.

5.10 Proposition Complete separability is hereditary.

5.11 Proposition A completely separable space is separable.

5.12 Proposition A separable metrisable space is completely separable.

The Arens–Fort space (a major example)

On the set $(\mathbb{N} \times \mathbb{N}) \cup \{(0,0)\}$, we define a topology τ_{af} by declaring:

$G \in \tau_{\text{af}} \Leftrightarrow$ **either** $(0,0) \notin G$
or $(0,0)$ does $\in G$ and G contains all but finitely many points in each of all but finitely many 'columns' $\{n\} \times \mathbb{N}$.

Verify that

(a) no sequence of elements of $\mathbb{N} \times \mathbb{N}$ can converge to $(0, 0)$,
(b) this space (although it is defined on a countable set!) is not first-countable,
(c) complete separability implies first-countability (and therefore the Arens–Fort space is not completely separable).

5.13 Notes

(i) The converse to 5.11 is false.
(ii) Separability is hereditary among metric spaces. Therefore the space in 4.42 is not metrisable.
(iii) Complete separability is not preserved by continuous surjections, but it is preserved by continuous open surjections.

Product spaces

5.14 Definitions Given a finite family

$$(X_1, \tau_1),\ (X_2, \tau_2),\ (X_3, \tau_3),\ \ldots\ ,\ (X_n, \tau_n)$$

of topological spaces, let

$$X\ =\ X_1\ \times\ X_2\ \times\ \ldots\ \times\ X_n\ =\ \prod_{i=1}^{n} X_i.$$

How may we best topologise X?
For any choice of a non-empty open set G_i in each X_i

$$G_1 \subseteq X_1,\ G_2 \subseteq X_2,\ G_3 \subseteq X_3,\ \ldots\ ,\ G_n \subseteq X_n,$$

the set

$$G_1\ \times\ G_2\ \times\ \ldots\ \times\ G_n\ =\ \prod_{i=1}^{n} G_i$$

is called an *open box*. Put

$$\tau\ =\ \{\emptyset \text{ and all unions of open boxes}\}.$$

Then τ is the *product topology* and (X, τ) the *product space* of the given family (Fig. 5.2).

5.15 Lemma In 5.14, τ is indeed a topology, and the family of all open boxes is a base for it.

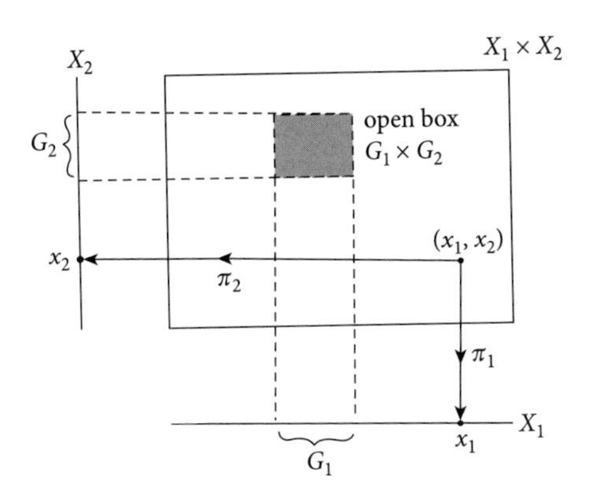

Fig. 5.2 Product of two spaces: projections and open boxes (see 5.14 and 5.17).

5.16 Examples It is readily checked that the product of two copies of the real line is identical to the coordinate plane under its usual topology. Indeed, the product of n copies of $(\mathbb{R}, \tau_{\text{usual}})$ is $\mathbb{R}^n$ with the topology derived from the natural Euclidean metric. This at least suggests that our definition, although long-winded, is reasonable!

5.17 Definition Continuing with the notation of 5.14, there are important mappings from X to the individual X_i's. For each i, define $\pi_i : X \to X_i$ by the formula

$$\pi_i((x_1, x_2, \ldots x_n)) = x_i.$$

This π_i is termed the ith *projection*. All it does is to pick out the ith coordinate of each element (the ith component of each vector, if you like) of the product set.

5.18 Lemma Each projection is a continuous open surjection.

5.19 Theorem A mapping (from any space) into a product space is continuous if and only if its composite with each projection is continuous (Fig. 5.3).

This result, 5.19, is so useful that advanced textbooks often take it as the *definition* of a product space! We next illustrate some of its applications.

5.20 Example For any spaces (X, τ) and (Y, τ'), $(X, \tau) \times (Y, \tau')$ is homeomorphic to $(Y, \tau') \times (X, \tau)$.

5.21 Example For any spaces (X, τ), (Y, τ') and (Z, τ''), the following two product spaces are homeomorphic:

$$[(X, \tau) \times (Y, \tau')] \times (Z, \tau''),$$
$$(X, \tau) \times [(Y, \tau') \times (Z, \tau'')].$$

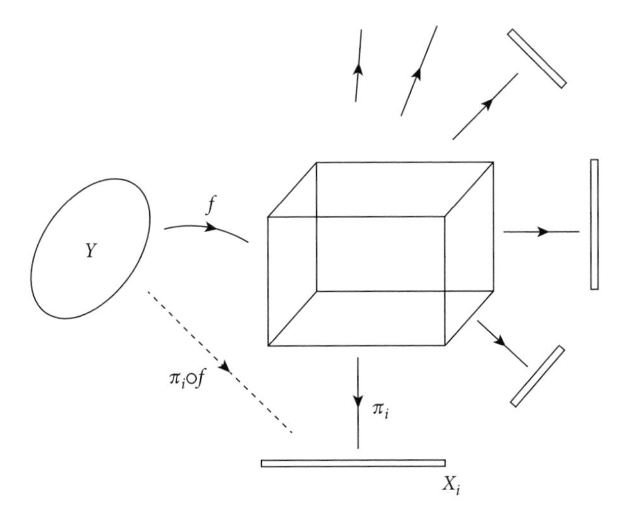

Fig. 5.3 Continuity of a map into a product space (see 5.19).

5.22 Note Essentially the same argument shows that for any finite list of spaces, 'the product of (the product of all except the last one) by the last one' is homeomorphic to the product of the complete list. The point of 5.20 and 5.21 is that we can build up any (finite) product by just multiplying two spaces at a time, and the order in which we do it is essentially irrelevant.

5.23 Example Choose an 'origin' (x_0, y_0) in $(X, \tau) \times (Y, \tau')$ and define $\eta : X \to X \times Y$ thus:

$$\eta(x) = (x, y_0).$$

Then the subspace $\eta(X)$ of $X \times Y$, the 'X-axis through the chosen origin', is homeomorphic to X.

Product theorems

There are a large number of theorems of the form 'Any product of NICE spaces is NICE also'. We exhibit a few of them.

5.24 Lemma If (X, τ) and (Y, τ') are sequentially compact, then so is their product.

5.25 Proposition The product of any finite number of sequentially compact spaces is sequentially compact.

5.26 Lemma Given subsets A of (X, τ_1) and B of (Y, τ_2), notice first that $A \times B$ is a subset of their product space $(X \times Y, \tau)$, say. Then

$$\overline{A \times B}^{\tau} = \overline{A}^{\tau_1} \times \overline{B}^{\tau_2}.$$

(Less formally but more memorably, the closure of a product is the product of the closures.)

5.27 Proposition The product of any finite number of separable spaces is separable.

5.28 Tychonoff's theorem – finite version The product of (any finite number of) compact spaces is compact.

It may not look very impressive at first sight, but (the infinite version of) 5.28 is arguably the most important theorem in topology. We give a small application:

5.29 Lemma The product of subspaces is a subspace of the product.

5.30 Example Any bounded closed subset of $(\mathbb{R}^n, d)$, where d is the usual Euclidean metric, is compact in the induced topology. This completes our evidence for 4.20!

Infinite products

Most of the difficulty in understanding products of infinitely many topological spaces has virtually nothing to do with topology – it lies in properly grasping what is meant by the product of an infinite family of sets. Look again at the finite case: if

$$(X_1, X_2, X_3, \ldots, X_n) = \{X_i : i \in \{1, 2, 3, \ldots, n\}\}$$

is a finite family of sets, then the typical point $(x_1, x_2, x_3, \ldots, x_n)$ in their product set is nothing but a way of spelling out, for each i in the labelling set $\{1, 2, 3, \ldots, n\}$ for the family, an object x_i that is somewhere in the union of all these sets and, to be more precise, has to belong to X_i, the ith member of the family. That is, an element in the product set is just a mapping x from $\{1, 2, 3, \ldots, n\}$ into $\bigcup_{i=1}^{n} X_i$ such that, for every i, we have $x(i) \in X_i$ (except that we usually write x_i rather than $x(i)$ in this context).

That looks like (and is!) a cumbersome way to describe the product set of a finite family, but the good news is that it transfers immediately to infinite families also:

5.31 Definitions

(i) Given an arbitrary family $\{X_i : i \in I\}$ of sets, their (Cartesian) *product* is the set of all maps x from the index set I into $\bigcup_{i \in I} X_i$ such that we have

$x(i) \in X_i$ for every $i \in I$. By convention, we continue to write $x(i)$ as x_i and to call it the ith *coordinate* of x.

(ii) Denote by $\prod_{i \in I} X_i$ this product set. For each $i \in I$, the ith *projection* is the mapping $\pi_i : \prod_{i \in I} X_i \to X_i$ given by the formula $\pi_i(x) = x_i$. So, exactly as in the finite case, it is the map that picks out the ith coordinate of each 'vector'.

5.32 Definition Given an arbitrary family $\{(X_i, \tau_i) : i \in I\}$ of topological spaces, consider the product set $X = \prod_{i \in I} X_i$ of their underlying sets. For any choice of $i \in I$ and non-empty $G_i \in \tau_i$, the set $\pi_i^{-1}(G_i)$ is called an *open cylinder* in X. The intersection of any finite collection of open cylinders is called an *open box* in X. There is a topology on X for which all the open cylinders form a subbase – and for which, therefore, all the open boxes form a base. It is called the *product topology*, and X endowed with this topology is called the *product space* of the given family (Fig. 5.4).

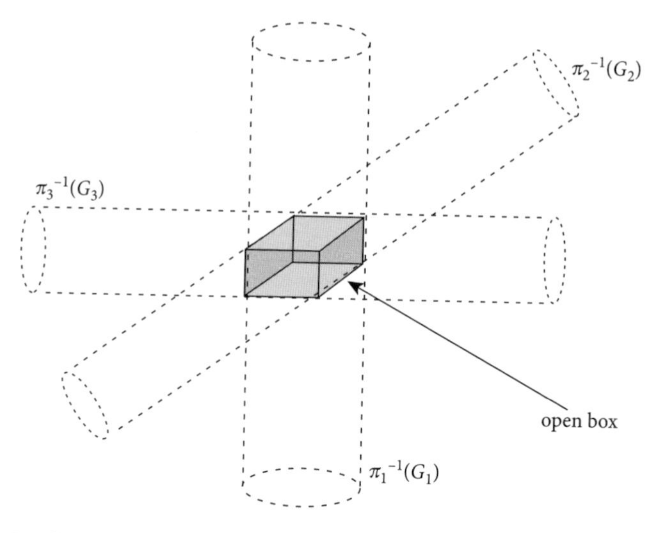

Fig. 5.4 Open cylinders and open boxes in an arbitrary product space (see 5.32).

5.33 Lemma Each projection is a continuous open surjection. Indeed, the product topology is the 'smallest' (in the natural sense) topology on the product set with respect to which all the projections are continuous.

5.34 Theorem A mapping (from any space) into a product space is continuous if and only if its composite with each projection is continuous.

Warning: not all of the product theorems we looked at in the previous section remain valid when we include *infinite* products. For instance, the product of an infinite number of sequentially compact spaces or of separable spaces may fail to enjoy the same property. But a very important fact is that Tychonoff's theorem remains valid for arbitrary products. The key step in (one of) the proof(s) is the following classy application of Zorn's lemma:

5.35 Alexander's subbase lemma Suppose that $\mathcal{S}$ is a subbase for a topological space (X, τ) and that every covering of X by members of $\mathcal{S}$ possesses a finite subcovering. Then (X, τ) is compact.

5.36 Tychonoff's theorem The product of an *arbitrary* family of compact topological spaces is compact.

Exercises Essential Exercises 44–46 and 48–57 are based on the material in this chapter. It is particularly recommended that you should try numbers 45, 48, 54, 55, 56 and 57.

Expansion of Chapter 5

5.3

(i) Suppose $\mathcal{B}$ is a base for τ.
Given $x \in G \in \tau$, it must be possible to write G in the form

$$G = \bigcup_{\alpha \in I} B_\alpha,$$

where each $B_\alpha \in \mathcal{B}$.
Choose $\alpha_0 \in I$ so that $x \in B_{\alpha_0}$.
Then $x \in B_{\alpha_0} \subseteq G$ and $B_{\alpha_0} \in \mathcal{B}$.

(ii) Suppose (1) holds.
Given a non-empty open set G, use (1) to choose, for each $x \in G$, some $B_x \in \mathcal{B}$ for which

$$x \in B_x \subseteq G.$$

Then $\bigcup_{x \in G} B_x$ is precisely G.
Hence $\mathcal{B}$ is a base for τ.

5.4 By 3.11, we may choose to work in base-10 arithmetic, since the Cantor sets from other bases are all homeomorphic to $Cantor_{10}$.
We recall that $Cantor_{10}$ is the intersection of the subsets of $\mathbb{R}$ that we denoted by $*^n[0, 1]$; each of these was the union of 2^n intervals, each of form $[p, p + 10^{-n}]$, and the gap from this interval to the nearest on either side was at least as long. Therefore

$$Cantor_{10} \cap [p, p + 10^{-n}] = Cantor_{10} \cap (p - 10^{-n}, p + 2 \times 10^{-n}).$$

By inspection, this set is both open and closed in $Cantor_{10}$.
Yet we saw (in 3.10) that in $Cantor_{10}$ every neighbourhood of an arbitrary point x contains a set of the above form $Cantor_{10} \cap [p, p + 10^{-n}]$ to which x belongs.
By 5.3, this is exactly what is needed to show that these sets form a base.

5.5 $f : (X, \tau) \to (Y, \tau'), \mathcal{B}$ a base for τ', $f^{-1}(B) \in \tau \; \forall B \in \mathcal{B}$.
Let $G \in \tau'$.
If $G = \emptyset$, then of course $f^{-1}(G) = \emptyset \in \tau$.
If $G \neq \emptyset$, then G is a union of some members of $\mathcal{B}$:

$$G = \bigcup_{\alpha \in I} B_\alpha \qquad (\text{each } B_\alpha \in \mathcal{B}).$$

Then

$$f^{-1}(G) = \bigcup_{\alpha \in I} f^{-1}(B_\alpha)$$

= a union of τ-open sets
therefore τ-open again.

Hence f is continuous.

5.6

(i) Suppose $p \in \overline{A}$.
By 2.14, every neighbourhood of p meets A.
In particular, whenever $p \in B \in \mathcal{B}$ then (because B is τ-open) B *is* a neighbourhood of p, and must meet A.

(ii) Suppose each member of $\mathcal{B}$ to which p belongs intersects A.
If N is any neighbourhood of p, then

choose open G s.t. $p \in G \subseteq N$,
choose $B \in \mathcal{B}$ s.t. $p \in B \subseteq G$ (5.3!).

Now B meets A, therefore so does G, therefore so does N.
Now 2.14 says $p \in \overline{A}$.

5.7

(i) Each set of the form $A \cap B$ (where $B \in \mathcal{B}$) is τ_A-open, since each $B \in \mathcal{B}$ is τ-open.

(ii) For each non-empty τ_A-open set G', choose $G \in \tau$ such that $G' = A \cap G$.
Now G is a union of certain members of $\mathcal{B}$:

$$G = \bigcup_{\alpha \in I} B_\alpha, \qquad \text{where each } B_\alpha \in \mathcal{B}.$$

Then

$$G' = A \cap \left(\bigcup_{\alpha \in I} B_\alpha \right) = \bigcup_{\alpha \in I} (A \cap B_\alpha).$$

By (i) and (ii), $\{A \cap B : B \in \mathcal{B}\}$ is a τ_A-base.

5.9

- $\{(q_1, q_2) : q_1 \in \mathbb{Q}, q_2 \in \mathbb{Q}, q_1 < q_2\}$ is a countable collection of open intervals and, via 5.3, it is a base for $(\mathbb{R}, \tau_{\text{usual}})$.

- Consider ι_p, an included-point topology on $\mathbb{R}$.
Let $\mathcal{B}$ be any base for it.
For each $x \neq p$ in $\mathbb{R}$, $\{p, x\}$ is an open set including x, so 5.3 tells us $\exists B_x \in \mathcal{B}$ such that

$$x \in B_x \subseteq \{p, x\}.$$

A little thought tells us $B_x = \{p, x\}$!
So $\mathcal{B}$ includes at least as many different members as there are points in $\mathbb{R} \setminus \{p\}$ and cannot be countable.

5.10 This is just about immediate from 5.7.

5.11 Given a countable base $\mathcal{B}$ for (X, τ),
for each non-empty B in $\mathcal{B}$ choose a point $x_B \in B$
and consider the countable set $D = \{x_B : B \in \mathcal{B}\}$.
For any $x \in X$ and neighbourhood N of x,
choose open G such that $x \in G \subseteq N$,
choose $B \in \mathcal{B}$ such that $x \in B \subseteq G$,
and we see that $x_B \in N$.
Thus N intersects D.

Thus $x \in \overline{D}$ via 2.14.
Thus $\overline{D}$ is the whole of X, that is, D is dense and (X, τ) is separable.

5.12 Suppose (M, d) is a separable metric space with countable dense C. Put

$$\mathcal{B} = \{B(c, q) : c \in C, q \in \mathbb{Q}, q > 0\}.$$

This is a countable family of open sets. We shall use 5.3 to show that it is a base.
Given $x \in G$, where G is open,
choose $\varepsilon > 0$ so that $B(x, \varepsilon) \subseteq G$.
Since C is dense, find $c \in C$ so that $d(x, c) < \frac{1}{3}\varepsilon$ (2.14 again).
Pick a rational q such that $\frac{1}{3}\varepsilon < q < \frac{2}{3}\varepsilon$.
Then $B(c, q)$ belongs to $\mathcal{B}$,

$$x \in B(c, q) \quad \text{since } q > \frac{1}{3}\varepsilon > d(x, c)$$

and

$$B(c, q) \subseteq B(x, \varepsilon) \text{ via the triangle inequality,}$$
$$\text{therefore } x \in B(c, q) \subseteq G \text{ and 5.3 applies.}$$

The Arens–Fort space

That's fairly hard to picture!
It says that, for G to be a neighbourhood of $(0, 0)$, there has to be a value of n such that, for every column to the 'right' of the nth, G must include all of that column except for a finite number of points. But every point except $(0, 0)$ is isolated, that is, is an open set on its own.
It is not entirely obvious that this even is a topology . . . but check it out.

(a) Let (x_n) be any sequence in $\mathbb{N} \times \mathbb{N}$.
If its terms are drawn from only a finite number of columns, then it is easy to find a neighbourhood of $(0, 0)$ that the sequence never gets inside, so $(x_n) \not\to (0, 0)$.

But also if (x_n) draws its terms from an infinite number of columns, we can extract a subsequence $(x_{n_1}, x_{n_2}, x_{n_3}, \ldots)$ that never includes two (or more) points in the same column. Then

$$\{(0,0)\} \cup (\mathbb{N} \times \mathbb{N}) \setminus \{x_{n_1}, x_{n_2}, x_{n_3}, \ldots\}$$

is a neighbourhood of $(0,0)$ which this subsequence never gets inside, so

$$(x_{n_k}) \not\to (0,0) \quad \text{and once again} \quad (x_n) \not\to (0,0).$$

(b) Yet $(0,0)$ does $\in \overline{\mathbb{N} \times \mathbb{N}}$ since each of its neighbourhoods meets the set (2.14 again).
Now (a) and 3.26 tell us that this space cannot be first-countable.

(c) For the proof that complete separability implies first-countability, see Essential Exercise 48(i).
This result and (b) tell us that the Arens–Fort space cannot be completely separable.

5.13

(i) The Arens–Fort space is separable – indeed, its entire underlying set is countable! – but not completely separable.
Alternatively, by 5.9, an included-point topology ι_p on $\mathbb{R}$ is not completely separable.
Yet the only ι_p-closed set that includes p is $\mathbb{R}$ itself,
so $\overline{\{p\}} = \mathbb{R}$
and $\{p\}$ is an (extremely!) countable dense subset,
establishing that $(\mathbb{R}, \iota_p)$ is separable.

(ii) For metric (and, therefore, for metrisable) spaces, separability is equivalent to complete separability (5.11, 5.12),
which is hereditary.
Now 4.42 shows a separable space with a non-separable subspace;
if it had been metrisable, the hereditariness of separability for metrisable spaces would have been contradicted.

(iii) For details of this proof, please see Essential Exercises 49 and 50 and their specimen solutions.

5.18 Consider the $i_0{}^{\text{th}}$ projection $\pi_{i_0} : \prod_{i=1}^{n} X_i \to X_{i_0}$.

- For a typical open box $B = G_1 \times G_2 \times \ldots \times G_n$,
  $\pi_{i_0}(B)$ is just G_{i_0}
  and is therefore τ_{i_0}-open.
  Since the open boxes form a base for the product topology, τ say,
  any τ-open set H is a union of open boxes

$$H = \bigcup_{\alpha \in I} B_\alpha$$

and

$$\pi_{i_0}(H) = \pi_{i_0}\left(\bigcup_\alpha B_\alpha\right) = \bigcup_\alpha \pi_{i_0}(B_\alpha)$$

= a union of τ_{i_0}-open sets, therefore τ_{i_0}-open itself.
Therefore π_{i_0} is an open map.

- For any $G_{i_0} \in \tau_{i_0}$,

$$\pi_{i_0}^{-1}(G_{i_0}) = X_1 \times X_2 \times \ldots \times X_{i_0-1} \times G_{i_0} \times \ldots \times X_n,$$

which is an open box!
Therefore it belongs to the usual base for τ,
therefore it belongs to τ.
Therefore π_{i_0} is continuous.

- Obviously, π_{i_0} is onto.

5.19 Consider $f : (Y, \tau') \to (X, \tau) = \prod_{i=1}^{n}(X_i, \tau_i)$.

(i) If f is continuous then recall that, for each i in the range 1 to n,
$\pi_i : X \to X_i$ is also continuous (5.18).
Therefore each of the $\pi_i \circ f$'s is also continuous (3.4).

(ii) Conversely, suppose that each $\pi_i \circ f$ is continuous.
For a typical open box $B = G_1 \times G_2 \times \ldots \times G_n$ in X, and typical $y \in Y$,

$$\begin{aligned} f(y) \in B &\Leftrightarrow (\pi_1 \circ f)(y) \in G_1 \;\&\; (\pi_2 \circ f)(y) \in G_2 \;\&\; \ldots \;\&\; (\pi_n \circ f)(y) \in G_n \\ &\Leftrightarrow y \in (\pi_1 \circ f)^{-1}(G_1) \cap (\pi_2 \circ f)^{-1}(G_2) \cap \ldots \cap (\pi_n \circ f)^{-1}(G_n), \end{aligned}$$

that is,

$$f^{-1}(B) = \bigcap_1^n (\pi_i \circ f)^{-1}(G_i)$$

$$= \text{a finite intersection of } \tau'\text{-open sets, therefore } \tau'\text{-open.}$$

Now 5.5 tells us that f is continuous.

5.20 Typical elements of $X \times Y$ and of $Y \times X$ look like

$$(x, y) \quad (y, x) \quad (x \in X, y \in Y).$$

Define $\theta : X \times Y \to Y \times X$ as shown in Fig. 5.5

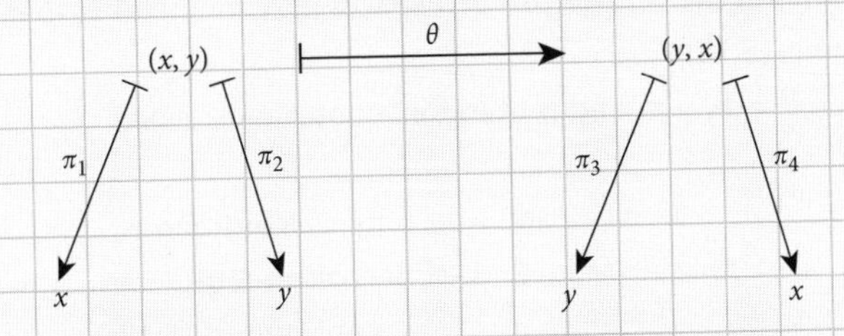

Fig. 5.5 Mapping $X \times Y$ to $Y \times X$ (see Expansion of 5.20).

– and label the projections. Now:

$$\pi_3 \circ \theta = \pi_2 \text{ is continuous (5.18),}$$
$$\pi_4 \circ \theta = \pi_1 \text{ is continuous (5.18),}$$
$$\text{therefore } \theta \text{ is continuous (5.19).}$$

Clearly $(y, x) \mapsto (x, y)$ is the inverse of (1–1, onto) θ, and the same method shows that this also is continuous.
So θ is a homeomorphism.

5.21 Be fussy about notation: typical point of first set is $((x, y), z)$, typical point of second set is $(x, (y, z))\ldots$ where $x \in X, y \in Y, z \in Z$.
Define θ from first set to second set as shown in Fig. 5.6

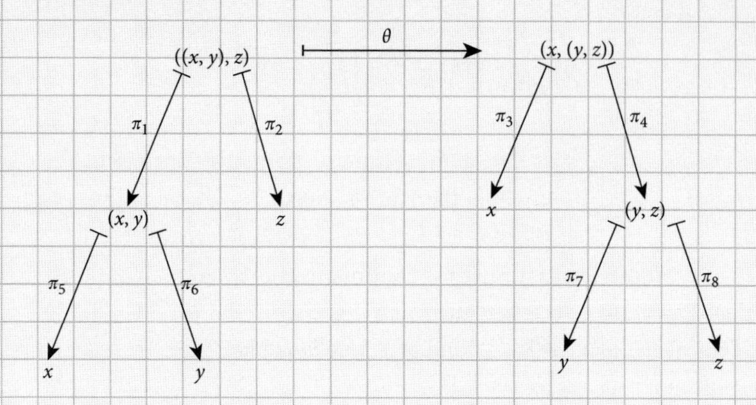

Fig. 5.6 Mapping $(X \times Y) \times Z$ to $X \times (Y \times Z)$ (see Expansion of 5.21).

– and we have also labelled the various projection maps that lift out the coordinates in the various product spaces here. Remember that they are all continuous (5.18).

Now $\pi_3 \circ \theta = \pi_5 \circ \pi_1$ and is therefore continuous; next:

$$\begin{aligned} \pi_7 \circ (\pi_4 \circ \theta) &= \pi_6 \circ \pi_1 \text{ is continuous,} \\ \pi_8 \circ (\pi_4 \circ \theta) &= \pi_2 \text{ is continuous,} \\ &\text{therefore } \pi_4 \circ \theta \text{ is continuous (5.19).} \end{aligned}$$

Since $\pi_3 \circ \theta$ and $\pi_4 \circ \theta$ are continuous, so is θ by 5.19.

Next, use the same kind of argument to show θ^{-1} is continuous. (Obvious that θ is 1–1 onto and therefore *has* an inverse.)

So θ is a homeomorphism.

5.23 See Fig. 5.7.

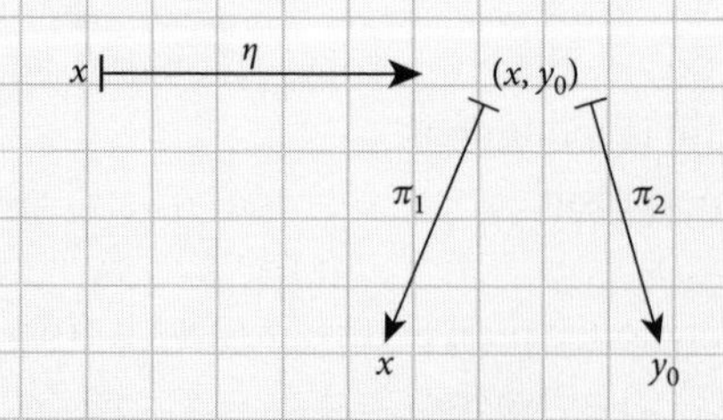

Fig. 5.7 Mapping X to a 'horizontal slice' through $X \times Y$ (see Expansion of 5.23).

$\pi_1 \circ \eta$ is id_X therefore continuous (3.9(i)),

$\pi_2 \circ \eta$ is constant therefore continuous (3.3(iii)).

Therefore η continuous (5.20).

Clearly, η is 1–1 (and onto its own range!)
and the inverse of $\eta : X \to X \times \{y_0\}$ is $(x, y_0) \mapsto x$,
that is, it is π_1 restricted to $X \times \{y_0\}$
and this is continuous by 5.18 and 3.5.
So X and $\eta(X)$ are homeomorphic (under η).

5.24 Let $((x_n, y_n))_{n\geq 1}$ be any sequence in $X \times Y$.
Then $(x_n)_{n\geq 1}$ is a sequence in (sequentially compact) X
so there exists a convergent subsequence $(x_{n_k})_{k\geq 1} \to$ some $\ell \in X$.
Next, $(y_{n_k})_{k\geq 1}$ is a sequence in (sequentially compact) Y
so there exists a convergent subsequence $(y_{n_{k_j}}) \to$ some $m \in Y$.
Also, $(x_{n_{k_j}})_{j\geq 1}$ is a subsequence of convergent (x_{n_k}).
Therefore $x_{n_{k_j}} \to \ell$ also.
It follows that

$$(x_{n_{k_j}}, y_{n_{k_j}}) \to (\ell, m) \text{ in } X \times Y,$$

and this is a subsequence of the original sequence.
So $X \times Y$ is sequentially compact.

(You should also check that
$(a_n, b_n) \to (p, q)$ in $X \times Y$ if and only if $a_n \to p$ in X and $b_n \to q$ in Y
because we have used that implicitly above.)

5.25 Let S_n be the statement 'The product of n sequentially compact spaces is sequentially compact.'
S_1 is trivial,
S_2 is 5.24.
Assume S_k is true for a particular $k \geq 2$.
Given $k + 1$ sequentially compact spaces $X_1, X_2, \ldots, X_{k+1}$,
we know $X_1 \times \ldots \times X_k$ is sequentially compact by assumption,
therefore $(X_1 \times \ldots \times X_k) \times X_{k+1}$ is sequentially compact by S_2.
But 5.23 tells us the latter space is homeomorphic to $X_1 \times \ldots \times X_k \times X_{k+1}$,
so that also is sequentially compact.
Thus S_{k+1} is also true.
By induction, all the statements S_n $(n \geq 1)$ are true.

5.26 Given any $(x, y) \in X \times Y$, we have:
$(x, y) \in \overline{A \times B} \Leftrightarrow$ every neighbourhood of (x, y) meets $A \times B$ (2.14)

$\Leftrightarrow$ every basic open neighbourhood of (x, y) meets $A \times B$ (5.3)
$\Leftrightarrow$ whenever $(x, y) \in$ open box $G \times H$, then $G \times H$ meets $A \times B$
$\Leftrightarrow$ whenever $x \in \tau_1$-open G and $y \in \tau_2$-open H, then G meets A and H meets B
(think!)
$\Leftrightarrow$ every neighbourhood of x meets A and every neighbourhood of y meets B
$\Leftrightarrow x \in \overline{A}$ and $y \in \overline{B}$
$\Leftrightarrow (x, y) \in \overline{A} \times \overline{B}$.

5.27 Given $(X_1, \tau_1), (X_2, \tau_2), \ldots, (X_n, \tau_n)$ separable,
choose countable $D_1 \subseteq X_1, D_2 \subseteq X_2, \ldots, D_n \subseteq X_n$ that are dense therein.
Then $D_1 \times D_2 \times \ldots \times D_n$ is countable within $X_1 \times X_2 \times \ldots \times X_n$
and $\overline{D_1 \times D_2 \times \ldots \times D_n} = \overline{D_1} \times \overline{D_2} \times \ldots \times \overline{D_n}$ (5.26 and induction!)
$= X_1 \times X_2 \times \ldots \times X_n$.

5.28 It is actually not much harder to prove the 'infinite' version! (See 5.36.) However, if you wish to read 5.35 (Alexander's subbase lemma) first, then the following application of it will establish 5.28 painlessly, and it should also make it easier for you to see (in 5.36) why the same technique proves the infinite version as well.
Let (X, τ) be the product of two compact spaces (X_1, τ_1) and (X_2, τ_2). The family of so-called open cylinders

$$\mathcal{G} = \{\pi_1^{-1}(G_1), \pi_2^{-1}(G_2) : \emptyset \neq G_1 \in \tau_1, \emptyset \neq G_2 \in \tau_2\}$$

is a subbase for the product topology τ on X, because each of these sets is an open box (for instance, $\pi_1^{-1}(G_1)$ is the box $G_1 \times X_2$) – and therefore an open set – and because each open box $G_1 \times G_2$ is the (finite) intersection of two of them: $\pi_1^{-1}(G_1) \cap \pi_2^{-1}(G_2)$ is exactly $G_1 \times G_2$. Let us call such a set as $\pi_1^{-1}(G_1)$ a 'vertical' open cylinder on the (rather tenuous) grounds that it lies parallel to the X_2-axis in the natural attempt to sketch what is happening, and such a set as $\pi_2^{-1}(G_2)$ a 'horizontal' open cylinder. The set named as G_1 or G_2 for one of these cylinders we shall call the 'foot' of the cylinder.

What Alexander's subbase lemma tells us is that, to prove (X, τ) compact, we only need to check that every cover of X **by elements of the subbase** $\mathcal{G}$ has a finite subcover. Assume, with a view to contradiction, that $\mathcal{T}$ is a covering of X by open cylinders that has no finite subcovering.

Now the feet of the vertical cylinders that belong to $\mathcal{T}$ cannot cover X_1: for if they did, the compactness of X_1 would imply the existence of a finite selection $\{G_1, G_2, \ldots, G_n\}$ of them that covered $X_1 \ldots$, in which case $\{\pi_1^{-1}(G_1), \pi_1^{-1}(G_2), \ldots, \pi_1^{-1}(G_n)\}$ would cover all of X using only finitely many members of $\mathcal{T}$, contrary to the assumption. Thus we can pick a point x_1 in X_1 that lies in no such foot.

Likewise, we can pick a point x_2 in X_2 that does not lie in the foot of any horizontal cylinder belonging to $\mathcal{T}$.

But the point (x_1, x_2) of X has to belong to some open cylinder in the cover $\mathcal{T}$ and, whether this is a horizontal one or a vertical one, we get an immediate contradiction to the way in which x_1 and x_2 were chosen.

This shows that the product of the two compact spaces must be compact. The usual inductive argument extends the conclusion to any (finite) number of spaces.

5.29 Consider the product (X, τ) of (X', τ') and (X'', τ'') and the product (A, τ^*) of subspaces $(A', \tau'_{A'}), (A'', \tau''_{A''})$, respectively.

- A typical basic ('open box') set in (A, τ^*) takes the form

$$(A' \cap G') \times (A'' \cap G'') \ldots \qquad (1)$$

where $G' \in \tau'$ and $G'' \in \tau''$.

- A typical basic set in (X, τ) takes the form $G' \times G''$.
- A typical basic set in *the subspace* (A, τ_A) of (X, τ) takes the form

$$\begin{aligned} &A \cap (G' \times G'') \\ &= (A' \times A'') \cap (G' \times G'') \ldots \end{aligned} \qquad (2)$$

Inspect (1) and (2) and realise that they are equal!
So τ^* and τ_A have the same basic sets, and are therefore the same topology.

5.33

- That $\pi_{i_0}: \prod_{i \in I}(X_i, \tau_i) \to (X_{i_0}, \tau_{i_0})$ is *onto* should be obvious.
- For any non-empty $G_{i_0} \in \tau_{i_0}, \pi_{i_0}^{-1}(G_{i_0})$ is an open cylinder, in the defining subbase for the product topology, therefore open.
So π_{i_0} is continuous.

- Points in a typical open box $\bigcap_{j=1}^{n} \pi_{i_j}^{-1}(G_{i_j})$ have their i_jth coordinates restricted to lie in G_{i_j} (for $1 \leq j \leq n$) but their *other* coordinates are free to range over the full space available. So

$$\pi_{i_0}\left(\bigcap_{1}^{n} \pi_{i_j}^{-1}(G_{i_j})\right) = \begin{cases} G_{i_0} & \text{if } i_0 \text{ equals one of the } i_j, \\ X_{i_0} & \text{otherwise.} \end{cases}$$

It follows that π_{i_0} maps each *open box* to an open set.
Then it follows as in 5.18 that π_{i_0} maps each *open set* in the product space to an open set,
that is, π_{i_0} is an open map.

- If σ is any topology on X that makes all of the projections continuous, then this continuity obliges each open cylinder $\pi_{i_j}^{-1}(G_{i_j})$ to be σ-open. Then each non-empty open set in the product topology, being formed by firstly 'finite-intersecting' and secondly 'unioning' these cylinders that are known to belong to the topology σ, must lie in σ also.

5.34

$$(Y, \tau') \longrightarrow (X, \tau) = \prod_{i \in I}(X_i, \tau_i).$$

- If f is continuous then so is every $\pi_i \circ f$, by 5.33 and 3.4.
- Conversely, suppose that every $\pi_i \circ f$ is continuous.
For a typical subbasic open cylinder $S = \pi_i^{-1}(G_i)$,

$$f^{-1}(S) = f^{-1}(\pi_i^{-1}(G_i)) = (\pi_i \circ f)^{-1}(G_i) \text{ is } \tau'\text{-open.}$$

The typical *basic* open box is a finite intersection of these:

$$S_1 \cap S_2 \cap \ldots \cap S_n,$$

so $f^{-1}(S_1 \cap \ldots \cap S_n) = \bigcap_1^n f^{-1}(S_i)$ is an intersection of finitely many τ'-open sets, therefore τ'-open.
Now 5.5 says f is continuous.

5.35 Proof: Assume (with a view to contradiction) that every cover from S has a finite subcover, but that X is not compact. Then X has an open cover without finite subcover. Let P = {all open covers of X which do not have finite subcovers}, which is now a non-empty collection; then $(P, \subseteq)$ is partially ordered with respect to inclusion. We shall use Zorn's lemma (which says that a partially ordered set in which every chain has an upper bound must have at least one maximal element).

Check for applicability of Zorn's lemma: Let Ç be a non-empty totally ordered subset of P. So Ç is a set of open covers in P such that $\forall \mathcal{T}_1, \mathcal{T}_2 \in$ Ç, either $\mathcal{T}_1 \subseteq \mathcal{T}_2$ or $\mathcal{T}_2 \subseteq \mathcal{T}_1$. Let $\bigcup$Ç = {all open sets which belong to at least one $\mathcal{T} \in$ Ç}. Certainly $\bigcup$Ç is an open cover of X. Take any finite subcollection $\{C_1, \ldots, C_n\}$. Each $C_i \in$ some $\mathcal{T}_i \in$ Ç; but since every two $\mathcal{T}_i$'s are 'comparable', there is a biggest $\mathcal{T}_j$ amongst $\mathcal{T}_1, \ldots, \mathcal{T}_n$. Then $\{C_1, \ldots, C_n\}$ is a finite subset of $\mathcal{T}_j \in P$, so cannot cover X. This shows that $\bigcup$Ç $\in P$, and certainly $\bigcup$Ç $\supseteq$ every $\mathcal{T}$ in Ç. That is to say, $\bigcup$Ç is an upper bound in P for the typical chain Ç in P. Therefore Zorn's lemma is applicable.

By Zorn's lemma, P has a maximal element $\mathcal{M}$. Thus:
(1) $\mathcal{M}$ is an open cover without finite subcover, but
(2) any bigger open cover must have a finite subcover.

We claim that $\mathcal{M} \cap \mathcal{S}$ is a cover of X.

Let $x \in X$. Then $\exists M \in \mathcal{M}$ such that $x \in M$, and since $\mathcal{S}$ is a subbase there are finitely many $S_1, \ldots, S_n \in \mathcal{S}$ such that $x \in \bigcap_{i=1}^{n} S_i \subseteq M$. If none of these $S_i \in \mathcal{M}$, then $\forall i\ \mathcal{M} \cup \{S_i\}$ has a finite subcover by (2):

$$X = S_i \cup (M_{i,1} \cup \ldots \cup M_{i,n_i}).$$

It follows that $X = \bigcap_{i=1}^{n} S_i \cup (\bigcup_{1 \le i \le n} \bigcup_{1 \le j \le n_i} M_{i,j})$, whence $X = M \cup \bigcup M_{i,j}$, which contradicts (1). So some S_i belongs to $\mathcal{M}$ as well as to $\mathcal{S}$. This proves that $\mathcal{M} \cap \mathcal{S}$ is a cover of X by elements from $\mathcal{S}$. By hypothesis $\mathcal{M} \cap \mathcal{S}$ has a finite subcover, which is then also a finite subcover from $\mathcal{M}$. This contradicts (1), finally, and we are done.

5.36 Let $(X, \tau) = \prod_{i \in I}(X_i, \tau_i)$, where $\{(X_i, \tau_i)\}_{i \in I}$ is a family of compact spaces with the product topology.
The family

$$\mathcal{G} = \{\pi_i^{-1}(G_i) : i \in I, \emptyset \neq G_i \text{ open in } X_i\}$$

is a subbase for the product topology τ on $\prod_{i \in I} X_i = X$. By Alexander's subbase lemma, we only need to check that every cover of X by elements from $\mathcal{G}$ has a finite subcover. Assume by way of contradiction that $\mathcal{T}$ is a covering of X by members of $\mathcal{G}$ that has no finite subcovering.

For each $i \in I$, we claim that $\{G_i \in \tau_i : \pi_i^{-1}(G_i) \in \mathcal{T}\}$ cannot cover X_i: because if it did, the compactness of X_i would imply that there exists a finite subcover of X_i, that is, $\exists G_{i_1}, \ldots, G_{i_n}$ in the family which cover X_i. Thus $\pi_i^{-1}(G_{i_1}), \ldots, \pi_i^{-1}(G_{i_n})$ would cover X and would belong to $\mathcal{T}$, which contradicts our assumption that $\mathcal{T}$ has *no* finite subcovering. Therefore $\exists p_i \in X_i$ such that p_i does not belong to any of the sets of the form $\{G_i \in \tau_i : \pi_i^{-1}(G_i) \in \mathcal{T}\}$.

Define $p = (p_i)_{i \in I} \in X$. Then p must belong to some member of $\mathcal{T}$, say $p \in \pi_i^{-1}(G_i) \in \mathcal{T}$. Thus $p_i = \pi_i(p) \in G_i$, which contradicts the choice of p_i. Thus we have the desired contradiction.

By Alexander's subbase lemma, X is compact.

6 Separation axioms

We have observed instances of topological statements which, although true for all metric (and all metrisable) spaces, fail for some topological spaces. Frequently, the cause of failure can be traced to there being 'not enough open sets' (in senses to be made precise). For instance, in any metric space, compact subsets are always closed; yet this is untrue in some topological spaces, because the proof rests ultimately on the observation

'given $x \neq y$, it is possible to find disjoint open sets G and H with $x \in G$ and $y \in H$',

which holds good in all metric spaces but fails in, for example, any trivial space.

Our programme in this chapter is to impose, upon the spaces that we study, successively stronger conditions, each of which 'demands certain minimum levels-of-supply of open sets', and to observe that in so doing we gradually eliminate the more pathological topologies, leaving us with those that behave like metric spaces to a greater or lesser extent. These conditions are known as the (classical) separation axioms.

T_1 spaces

6.1 Definition We call (X, τ) T_1 if, for each $x \in X$, $\{x\}$ is closed.

6.2 Comments

(i) Every metrisable space is T_1.

(ii) Trivial spaces are not T_1 (except when the underlying set is a singleton).

(iii) T_1 implies that every finite set is closed; indeed, the converse is also valid.

(iv) Any product of T_1 spaces is T_1.

(v) The property T_1 is hereditary.

The respects in which T_1 spaces behave better than unrestricted spaces are largely concerned with the idea of *cluster point of a set* (which we have not needed to use). We note the equivalence, in T_1 spaces, of the two forms of its definition that turn up frequently in analysis:

6.3 Proposition Given a T_1 space (X, τ), a point p of X and a subset A of X, the following are equivalent:

(i) each neighbourhood of p contains infinitely many points of A,
(ii) each neighbourhood of p contains a point of A different from p.

T_2 spaces

6.4 Definition We call (X, τ) T_2 or *Hausdorff* if:

for each distinct $x, y \in X$, $\exists G, H \in \tau$ such that $x \in G$, $y \in H$, $G \cap H = \emptyset$

(Fig. 6.1).

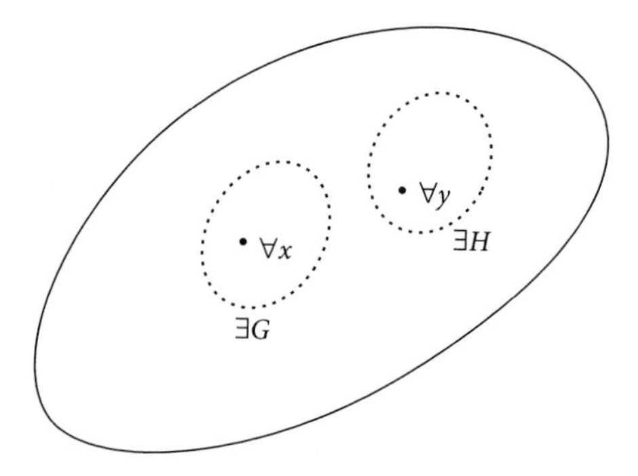

Fig. 6.1 The T_2 property (see 6.4).

6.5 Comments

(i) Every metrisable space is T_2.
(ii) Every T_2 space is T_1.
(iii) For any infinite X, (X, τ_{cf}) is T_1 but not T_2.
(iv) The property T_2 is hereditary.
(v) Any product of T_2 spaces is T_2.

The T_2 axiom is particularly valuable when exploring compactness, partly because although it only demands that there be 'enough open sets to separate points', it follows from this that there are enough to separate points from compact sets, *and even* to separate one compact set from another, in the same sense.

6.6 Proposition Within a T_2 space (X, τ), if C is compact and x does not belong to C, then $\exists G, H \in \tau$ such that $x \in G$, $C \subseteq H$, $G \cap H = \emptyset$.

6.7 Corollary (1) Compact subsets of T_2 spaces are closed.

6.8 Corollary (2) Within a T_2 space (X, τ), if C and K are compact and disjoint, then $\exists G, H \in \tau$ such that $K \subseteq G$, $C \subseteq H$, $G \cap H = \emptyset$ (Fig. 6.2).

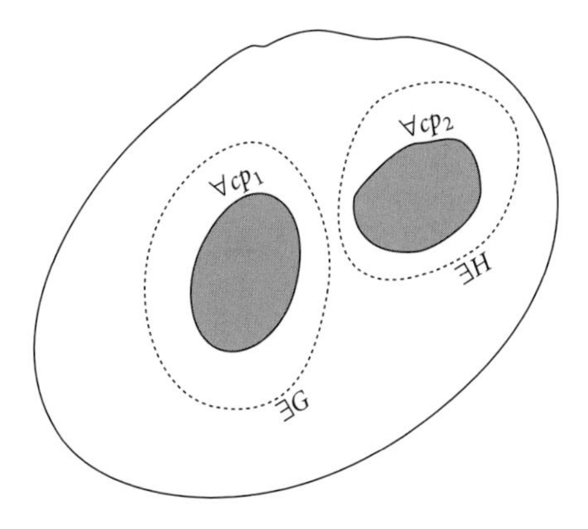

Fig. 6.2 Separating compact sets in a T_2 space (see 6.8).

A basic formal distinction between 'most algebra' and 'most topology' is that, whereas the inverse of a one-to-one onto group (etc!) homomorphism is automatically a homomorphism, the inverse of a one-to-one onto continuous map may very well fail to be continuous. It is an interesting consequence of 6.7 that, amongst compact T_2 spaces, this cannot happen.

6.9 Proposition Suppose that $f:(X,\tau)\to(Y,\tau')$ is one-to-one and onto and continuous, where (X,τ) is compact and (Y,τ') is T_2. Then f is a homeomorphism.

6.10 Example

(i) Let $2^{\mathbb{N}}$ denote the product of a countably infinite family of spaces each of which is a two-point set – say, $\{0, 1\}$ – with discrete topology. Then this space is homeomorphic to the Cantor set.

(ii) The product of finitely or of countably infinitely many copies of the Cantor set is homeomorphic to the Cantor set.

The T_2 property relates (although not especially well) to uniqueness of limits for sequences:

6.11 Proposition If (X,τ) is T_2, then no sequence in X can have more than one limit.

6.12 Note The converse of 6.11 fails.

Once again, nets are better than sequences at describing how topological spaces operate:

6.13 Proposition A topological space is T_2 if and only if none of its nets possesses two or more different limits.

T_3 spaces

6.14 Definition We call (X, τ) T_3 or *regular* if

(i) it is T_1 and

(ii) for each closed subset F of X and each point $x \in X \setminus F$, $\exists G, H \in \tau$ such that $x \in G$, $F \subseteq H$, $G \cap H = \emptyset$ (Fig. 6.3).

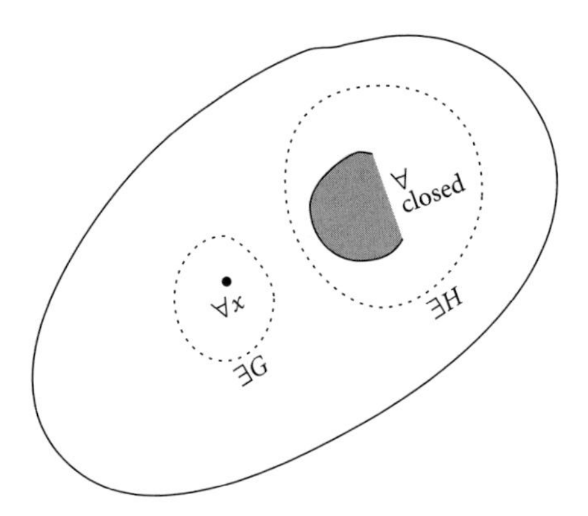

Fig. 6.3 The regularity property (see 6.14).

6.15 Lemma A T_1 space (X, τ) is $T_3 \Leftrightarrow$:

$$\text{given } x \in \text{open } G, \exists \text{ open } H \text{ such that } x \in H \subseteq \overline{H} \subseteq G. \qquad (*)$$

6.16 Comments

(i) This is not an especially important separation axiom for our purposes, so we shall merely fit it into the logical hierarchy.

(ii) Every metrisable space is T_3.

(iii) Every T_3 space is T_2.

(iv) One can devise examples of T_2 spaces that are not T_3.

(v) The T_3 condition is both productive (that is, preserved by forming products) and hereditary.

(vi) *Warning:* some writers take T_3 to mean 6.14 part (ii) alone, and 'regular' to mean 6.14 both (i) and (ii). Other writers do exactly the opposite! *Caveat lector* ...

$T_{3\frac{1}{2}}$ spaces

6.17 Definition We call (X, τ) $T_{3\frac{1}{2}}$ or *completely regular* or *Tychonoff* if

(i) it is T_1 and

(ii) for each closed (non-empty) subset F of X and each point $x \in X \setminus F$,

$$\exists \text{ continuous } f : X \to [0,1] \text{ such that } f(x) = 1 \text{ and } f(F) = \{0\}$$

(Fig. 6.4).

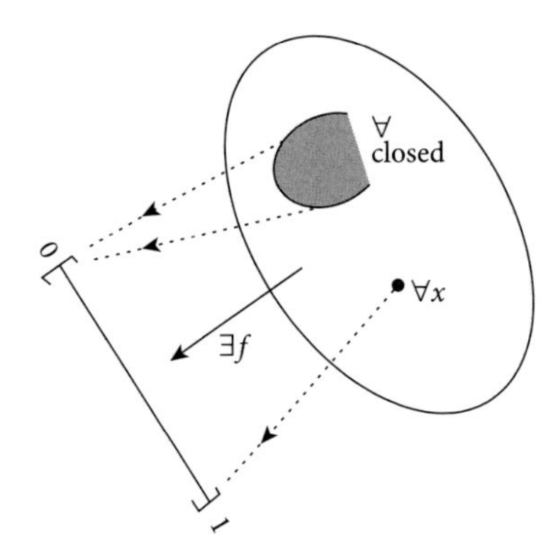

Fig. 6.4 The complete regularity (or Tychonoff) property (see 6.17).

6.18 Comments

(i) In condition (ii) here, we could equally well have said $f(x) = 0$ and $f(F) = \{1\}$, because if f were to satisfy this condition, then the real function f' given by the formula $f'(x) = 1 - f(x)$ would be continuous, and would separate x from F in the way that (ii) required (and vice versa).

(ii) Every metrisable space is Tychonoff.

(iii) Every Tychonoff space is T_3.

(iv) One can devise examples of T_3 spaces that are not Tychonoff.

(v) The Tychonoff property is productive and hereditary.

(vi) *Warning:* there is again some risk of confusion in the literature as to whether T_1 is or is not included in the meaning of $T_{3\frac{1}{2}}$/completely regular/Tychonoff.

T_4 spaces

6.19 Definition We call (X, τ) T_4 or *normal* if

(i) it is T_1 and

(ii) for each two disjoint closed (non-empty) subsets F_0 and F_1 of X,

$$\exists G, H \in \tau \text{ such that } F_0 \subseteq G,\ F_1 \subseteq H,\ G \cap H = \emptyset$$

(Fig. 6.5).

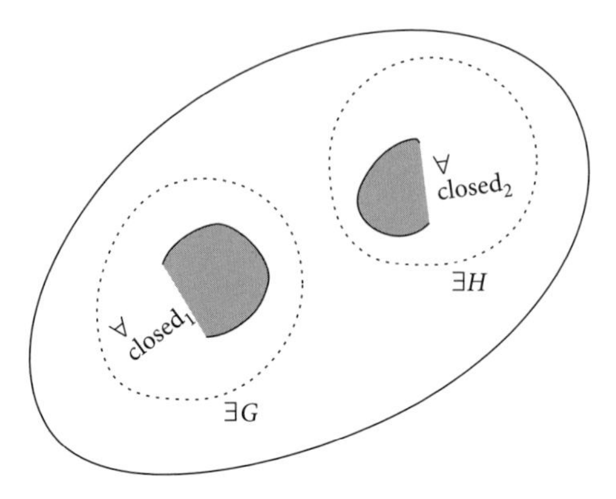

Fig. 6.5 The normality property (see 6.19).

6.20 Proposition Every metrisable space is normal.

It is true that normal implies Tychonoff, but it is not obvious! First, note that if G_0 and G_1 are open sets in a normal space such that $\overline{G_0} \subseteq G_1$, then we can find an open set $G_{\frac{1}{2}}$ such that $\overline{G_0} \subseteq G_{\frac{1}{2}}$ and $\overline{G_{\frac{1}{2}}} \subseteq G_1$. This is the first step towards a proof of:

6.21 Urysohn's lemma Let F_0 and F_1 be two disjoint closed (non-empty) subsets of a normal space (X, τ). Then there exists a continuous function $f : X \to [0, 1]$ such that

$$f(F_0) = \{0\} \text{ and } f(F_1) = \{1\}$$

(Fig. 6.6).

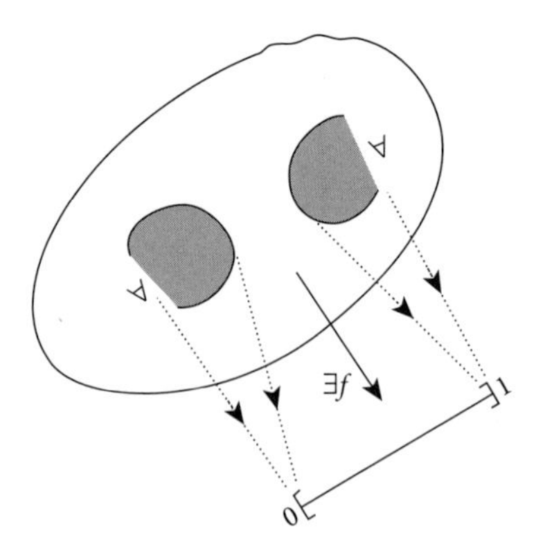

Fig. 6.6 Urysohn's lemma (see 6.21).

6.22 Corollary Every normal space is Tychonoff.

Examples are known of Tychonoff spaces that are not normal: see 6.30.

Compactness greatly simplifies the study of separation axioms. For instance:

6.23 Proposition A compact T_2 space is normal.

6.24 Note Unlike the properties previously considered in the 'hierarchy' of this chapter, normality is neither hereditary (see 6.30) nor productive.

The following is offered as an indication of how near we are to having come 'full circle':

6.25 Theorem Any completely separable, normal space is metrisable.

6.26 Definition Any chain (that is, totally ordered set) $C = (C, \leq)$ can be given a topology in a natural way, by using its open intervals exactly as one does in the real line to form τ_{usual} there. 'Bounded' and 'semi-infinite' open intervals in C are defined just as one might expect: for instance, $(a, b) = \{x \in C : a < x < b\}$ and $(a, \ \ ) = \{x \in C : a < x\}$. Then the **intrinsic topology** $\tau(\leq)$ on C is the topology that has the family of all the open intervals as its base.

6.27 Notes

(i) If this procedure is carried out on the real line, or on an interval on the real line, or on the rational line (all ordered in the usual manner), it produces exactly the usual topology that these sets carry.

(ii) More interesting things happen if you do it on an ordinal number: see below.

(iii) Since every 'bounded' interval (a, b) is the intersection of exactly two semi-infinite intervals $(a, \ \ )$ and $(\ \ , b)$, the semi-infinite intervals form a subbase for the intrinsic topology.

(iv) Intrinsic topologies are always T_2.

6.28 Lemma A well-ordered set under its intrinsic topology is compact if and only if it possesses a maximum element.

6.29 Example (the second uncountable ordinal) Recall the 'first uncountable ordinal'. This is a well-ordered set that is uncountable **but** all of whose initial segments are countable. Each of its countable subsets is bounded above. It has a least element (of course), which we shall denote by 0, but no maximum element. If we attach a maximum element t, we create a new well-ordered set W which is, in effect, the second uncountable ordinal. By 6.27(iv) and 6.28 it is compact and T_2 under its intrinsic topology, and therefore also T_4: see 6.23.

(i) Although t is in the closure of the segment $(\ \ , t)$, there is no sequence in $(\ \ , t)$ that converges to t. In particular, therefore, W cannot be first-countable.

(ii) Of course, there has to be a net in (, t) that does converge to t. Can you see the most natural example of one?

6.30 Example (the Tychonoff plank) Let us consider the following set Y of real numbers:

$$\{1 - 1/n : n \in \mathbb{N}\} \cup \{1\}.$$

In its natural order, it is a well-ordered set – indeed, it is effectively the second infinite ordinal! – but more relevantly it is compact, T_2 and T_4 for the same reasons as in 6.29. By Tychonoff's theorem (5.36), the product space $W \times Y$ is also compact, and T_2 via 6.27 and 6.5, and again T_4. The subspace $(W \times Y)\setminus\{(t, 1)\}$, that is, the entire space with its top right-hand corner point removed, is called the Tychonoff plank. Let us temporarily denote it by TP.

(i) TP is $T_{3\frac{1}{2}}$.

(ii) TP is not T_4.

(iii) So $T_{3\frac{1}{2}}$ does not imply T_4: see 6.22.

(iv) Also, T_4 is not hereditary: see 6.24.

Exercises Essential Exercises 58–79 are based on the material in this chapter. It is particularly recommended that you should try numbers 58, 60, 61, 62, 64, 65, 68 and 73.

Expansion of Chapter 6

6.2

(i) If (X, τ) is metrisable, that is, $\tau = \tau_d$ for some metric d, and $x \in X$, then, for each $y \neq x$, $B(y, d(x, y))$ excludes x but includes y. So

$$\bigcup_{y \neq x} B(y, d(x, y)) = X \setminus \{x\}$$

is open (being a union of open sets!), so $\{x\}$ is closed.

(iii) 'Each finite set is closed' $\Rightarrow$ 'Each singleton is closed'.
'Each singleton is closed' $\Rightarrow$ each finite set is a finite union of singletons = a finite union of closed sets and is therefore closed.

(iv) Let $(X, \tau) = \prod_{i \in I}(X_i, \tau_i)$ where each (X_i, τ_i), is T_1. For each $x \in X$:

$$\{x\} = \prod_{i \in I}\{x_i\} \quad \text{[think!]}$$

$$\text{therefore} \quad \overline{\{x\}} = \prod_{i \in I} \overline{\{x_i\}} \quad (5.26)^*$$

$$= \prod_{i \in I} \{x_i\} \quad \text{since each } (X_i, \tau_i) \text{ is } T_1$$

$$= \{x\}.$$

So $\{x\}$ is closed and (X, τ) is T_1.
*To be precise, this is not 5.26, but rather the 'infinite version' of 5.26. However, this is proved in the same way.

Alternatively, let x be a point in the product $(X, \tau) = \prod_{i \in I}(X_i, \tau_i)$, where each (X_i, τ_i) is T_1.

For each $y \neq x$ choose i_0 so that $x_{i_0} \neq y_{i_0}$.
In (X_{i_0}, τ_{i_0}), observe that $X_{i_0} \backslash \{x_{i_0}\}$ is τ_{i_0}-open and includes y_{i_0}.
Then $\pi_{i_0}^{-1}(X_{i_0} \backslash \{x_{i_0}\})$ is an open cylinder, therefore open in τ, and it includes y but excludes x.
Thus $X \setminus \{x\}$ is a τ-neighbourhood of every one of its own points, and is therefore open.
Consequently $\{x\}$ is closed, and (X, τ) is T_1.

(v) If (A, τ_A) is a subspace of a T_1 space (X, τ)
and $a \in A$
then $\{a\}$ is τ-closed
so $\{a\} = A \cap \{a\}$ is τ_A-closed (2.16),
therefore (A, τ_A) is T_1.

6.3

(a) (i) $\Rightarrow$ (ii) is trivially true.

(b) Suppose (ii) true.
If it were possible for some neighbourhood N of p to include only finitely many elements $a_1, a_2, \ldots, a_n$ of A distinct from p, then

$$N \cap (X \setminus \{a_1, a_2, \ldots, a_n\}) \text{ is a neighbourhood of } p$$

(because the expression in parentheses is an open set including p)

and it includes NO elements of A distinct from p, contradiction to (ii).
So each neighbourhood of p has to include infinitely many elements of A.

6.5

(i) In metrisable $(X, \tau) = (X, \tau_d)$ for some metric d,
if $x \neq y$, put $\varepsilon = \frac{1}{2} d(x, y)$
and we see $B(x, \varepsilon)$ and $B(y, \varepsilon)$ are disjoint open neighbourhoods of x and y.

(ii) If (X, τ) is T_2
and $x \in X$
then for each $y \neq x$ we can choose disjoint open G_x, H_x such that $x \in G_x$ and $y \in H_x$.
So $\bigcup_{y \neq x} H_x$ is exactly $X \setminus \{x\}$.
Then $X \setminus \{x\}$ is open and $\{x\}$ is closed,
therefore (X, τ) is T_1.

(iii) Immediate that (X, τ_{cf}) is T_1.
If (X, τ_{cf}) is T_2 and X is infinite,
choose $x \neq y$ in X
choose disjoint open G, H with $x \in G$ and $y \in H$. Then

$$X = X \setminus \emptyset = X \setminus (G \cap H) = (X \setminus G) \cup (X \setminus H)$$
$$= \text{the union of two finite sets}$$
$$\text{therefore finite } \ldots \textit{ contradiction!}$$

(iv) If (A, τ_A) is a subspace of a T_2 space (X, τ)
and $a \neq a'$ in A
then $a \neq a'$ in X
so $\exists$ disjoint τ-open G, H with $a \in G$ and $a' \in H$.
So $a \in A \cap G \in \tau_A$, $a' \in A \cap H \in \tau_A$
and $(A \cap G) \cap (A \cap H) = A \cap (G \cap H) = \emptyset$
so (A, τ_A) is T_2.

(v) Let $x \neq y$ in $(X, \tau) = \prod_{i \in I} (X_i, \tau_i)$, where each (X_i, τ_i) is T_2.
Choose i_0 so that $x_{i_0} \neq y_{i_0}$.
In (X_{i_0}, τ_{i_0}), choose disjoint open G, H such that $x_{i_0} \in G, y_{i_0} \in H$.
Then $\pi_{i_0}^{-1}(G), \pi_{i_0}^{-1}(H)$ are disjoint, contain x, y, respectively, and are open cylinders, therefore open in τ.
So (X, τ) is T_2.

6.6 For each $y \in C$, choose disjoint open G_y, H_y such that $x \in G_y, y \in H_y$.
The family $\{H_y: y \in C\}$ of open sets covers compact C, so there is a finite subcover:

$$C \subseteq \bigcup_1^n H_{y_i} = H \text{ (say).}$$

Also, $x \in \bigcap_1^n G_{y_i} = G$, which is open, since it is a *finite* intersection!
But $G \cap H = \emptyset$.

6.7 In the notation of 6.6, $x \in G \subseteq X \setminus H \subseteq X \setminus C$,
so $X \setminus C$ is a neighbourhood of every point of $X \setminus C$.
Therefore $X \setminus C$ is open (2.6).
Therefore C is closed.

6.8 For each $x \in K$, use 6.6 to find disjoint open G_x, H_x such that

$$x \in G_x, C \subseteq H_x.$$

Now $\{G_x: x \in K\}$ is an open cover of compact K, so $\exists$ finite subcover:

$$K \subseteq \bigcup_1^n G_{x_i} = G, \text{ say,}$$

$$\text{and } C \subseteq \bigcap_1^n H_{x_i} = H \text{ (say) (\textit{open} because \textit{finite} intersection).}$$

Now $G \cap H = \emptyset$.

6.9 If K is closed in X
then K is compact (4.14).
Therefore $f(K)$ is compact in Y (4.15).
Therefore $f(K)$ is closed (6.7).
This shows that f is a closed map.
Now 3.15 says f is a homeomorphism.

6.10

(i) Typical element of $2^{\mathbb{N}}$: a sequence $(x_n)_{n\geq 1}$ such that each x_n is either 0 or 1.
Typical element of $Cantor_{10}$: a real number expressible as decimal $0.d_1d_2d_3d_4\cdots$ such that each d_n is either 0 or 9.

Map these two sets to one another in the more-or-less obvious manner, namely
$f : 2^{\mathbb{N}} \to Cantor_{10}$ and $g : Cantor_{10} \to 2^{\mathbb{N}}$, described by

$$f((x_n)_{n\geq 1}) = 0.(9x_1)(9x_2)(9x_3)(9x_4)\cdots,$$
$$g(0.d_1d_2d_3d_4\cdots) = (d_n/9)_{n\in\mathbb{N}}.$$

Since f and g are mutually inverse and $2^{\mathbb{N}}$ is compact by Tychonoff's theorem and the metric space $Cantor_{10}$ is T_2, 6.9 tells us that it is now enough to prove that f is continuous.
If r is a positive integer, then $\bigcap_{i=1}^{r} \pi_i^{-1}(\{x_i\})$ is a neighbourhood of $2^{\mathbb{N}}$'s typical element $x = (x_n)_{n\geq 1}$, every element of which has the same first r coordinates as x has. Therefore, if $y = (y_n)_{n\geq 1}$ belongs to this neighbourhood, we note that $|f(x) - f(y)|$ is a decimal whose first r decimal places are zero: that is, $|f(x) - f(y)| < 10^{-r}$. Since 10^{-r} can be made arbitrarily small, this shows f to be continuous.

(ii) A finite or countably infinite product of copies of $2^{\mathbb{N}}$ is still just a product of a countable infinity of two-element discrete spaces.

6.11 In a T_2 space (X, τ), suppose sequence $(x_n) \to$ both ℓ and ℓ', where $\ell \neq \ell'$. Choose disjoint open neighbourhoods G of ℓ, H of ℓ'.

$$\exists n_0 \quad \text{such that} \quad n \geq n_0 \Rightarrow x_n \in G,$$
$$\exists n_1 \quad \text{such that} \quad n \geq n_1 \Rightarrow x_n \in H.$$

Pick $n_2 \geq$ both n_0 and n_1, and we get $x_{n_2} \in G \cap H = \emptyset$ – which is nonsensical!

6.12 For instance, we saw in 3.17(v) that in $(\mathbb{R}, \tau_{cc})$ we have $x_n \to \ell \Leftrightarrow x_n = \ell$ for all sufficiently large n, so certainly no sequence can converge to two different limits here. Yet $(\mathbb{R}, \tau_{cc})$ is not T_2: proof just like that of 6.5(iii).

6.13

(i) If (X, τ) is T_2, then 'uniqueness of limits for nets' is proved on very much the same lines as 6.11.

(ii) If (X, τ) is *not* T_2, we shall construct a net possessing two different limits.

There must exist x, y such that $x \neq y$ but every open set including x intersects every open set including y ... and therefore every neighbourhood of x intersects every neighbourhood of y.
Put an order $\leq$ on the set $\mathcal{N}_x \times \mathcal{N}_y$ like this:

$$(N_x, N_y) \leq (N'_x, N'_y) \Leftrightarrow \text{ both } N_x \supseteq N'_x \text{ and } N_y \supseteq N'_y.$$

Easy to check that $\mathcal{N}_x \times \mathcal{N}_y$ is now a directed set (see, for instance, Essential Exercise 24(i)).
For each $(A, B) \in \mathcal{N}_x \times \mathcal{N}_y$, pick a point $x_{(A,B)} \in A \cap B$.
Then $\left(x_{(A,B)}\right)_{(A,B) \in \mathcal{N}_x \times \mathcal{N}_y}$ is a net of points of X.

For any neighbourhood N of x, $(N, X) \in \mathcal{N}_x \times \mathcal{N}_y$ and:

$$\begin{aligned}(A, B) \geq (N, X) &\Rightarrow A \subseteq N \\ &\Rightarrow x_{(A,B)} \in A \cap B \subseteq A \subseteq N.\end{aligned}$$

Therefore this net converges to x.
Similarly, it converges to y.

6.15

(i) If X is T_3 and we are given $x \in$ open G, then $x \notin$ closed $X \setminus G$; so $\exists$ disjoint open H, J such that $x \in H, X \setminus G \subseteq J$.
Then $H \subseteq X \setminus J$ so $\overline{H} \subseteq \overline{X \setminus J} = X \setminus J \subseteq X \setminus (X \setminus G) = G$:
so $(*)$ is satisfied.

(ii) The converse '$(*) \Rightarrow T_3$' is very similar.

6.16

(ii)

- We already know that metrisable spaces are T_1.

- Given point x and closed set F in metrisable $(X, \tau) = (X, \tau_d)$, where d is a metric, and $x \notin F$:

$$x \in \text{ open } X \setminus F \text{ so } \exists \varepsilon > 0 \text{ such that } B(x, \varepsilon) \subseteq X \setminus F.$$

Put $G = B(x, \frac{1}{2}\varepsilon)$ and $H = \{y \in X : d(x, y) > \frac{1}{2}\varepsilon\}$
and we get the separation property that 6.14(ii) asks for.

(iii) If (X, τ) is T_3 then

- it is T_1 and
- given $x \neq y$ in X, we have $x \notin$ closed $\{y\}$! So 6.13 tells us $\exists$ disjoint open G and H such that

$$x \in G \quad \text{and} \quad \{y\} \subseteq H$$

– which is just an eccentric way of expressing T_2.

(v) In a product $(X, \tau) = \prod_{i \in I}(X_i, \tau_i)$ of T_3 spaces, let $x \notin$ closed F.
Through 5.3, there is a basic open box $B = \bigcap_{j=1}^{n} \pi_{i_j}^{-1}(G_{i_j})$, where each G_{i_j} is τ_{i_j}-open, such that $x \in B \subseteq X \setminus F$.
For each $j = 1, 2, \ldots, n$:
$x_{i_j} \in G_{i_j}$ (open)
so $\exists$ open H_{i_j} such that $x_{i_j} \in H_{i_j} \subseteq \overline{H}_{i_j} \subseteq G_{i_j}$ (see 6.15).
Now $\bigcap_{j=1}^{n} \pi_{i_j}^{-1}(H_{i_j}) = B_0$ is a new (reduced) open box including x and contained in the closed set

$$\bigcap_{j=1}^{n} \pi_{i_j}^{-1}(\overline{H}_{i_j}) \subseteq B,$$

and so $\overline{B}_0 \subseteq B \subseteq X \setminus F$,
that is, B_0 and $X \setminus \overline{B}_0$ wrap x and F in disjoint open sets.

Of course, (X, τ) is also T_1 (6.2(iv))
so it is T_3.

If (A, τ_A) is a subspace of a T_3 space (X, τ), we already know from 6.2(v) that A is T_1.
Given $a \in A, B \subseteq A, a \notin B, B$ being τ_A-closed:
we know $a \notin \overline{B}^{\tau_A} = A \cap \overline{B}^{\tau}$ (2.18).
Therefore $a \notin \overline{B}^{\tau}$, which is τ-closed.
Can choose disjoint τ-open G, H such that

$$a \in G, \overline{B}^{\tau} \subseteq H.$$

Now $A \cap G, A \cap H$ are disjoint τ_A-open, and

$$a \in A \cap G, A \cap H \supseteq A \cap \overline{B}^{\tau} = \overline{B}^{\tau_A} = B.$$

So (A, τ_A) is T_3.

6.18

(ii) It will be simpler to deduce that from 6.20 and 6.22.

(iii) Let (X, τ) be $T_{3\frac{1}{2}}$.
It is T_1, of course.
Given $x \notin$ closed F, we can find continuous $f: X \to [0,1]$ such that $f(x) = 1$ and $f(F) = \{0\}$.
Then $G = f^{-1}\left((\frac{1}{2}, 1]\right), H = f^{-1}\left([0, \frac{1}{2})\right)$ are open and disjoint, and wrap x and F in the way required for T_3.

(v) If $(X, \tau) = \prod_{i \in I}(X_i, \tau_i)$ is a product of $T_{3\frac{1}{2}}$ spaces, then we already know that it is T_1 (6.2(iv)).
Given $x \notin$ closed F within (X, τ),
find a basic open box $B = \bigcap_1^n \pi_{i_j}^{-1}(G_{i_j})$ such that

$$x \in B \subseteq X \setminus F.$$

For each $j = 1, 2, \ldots, n$:

$$x_{i_j} \in G_{i_j} \text{ (open) therefore } x_{i_j} \notin \text{closed } X_{i_j} \setminus G_{i_j},$$

so there is a continuous function

$$f_{i_j}: X_{i_j} \to [0,1] \text{ such that } \begin{cases} f_{i_j}(x_{i_j}) & = 1, \\ f_{i_j}(X_{i_j} \setminus G_{i_j}) & = \{0\}. \end{cases}$$

We define $f : X \to [0,1]$ thus:

$$f(y) = f_{i_1}(y_{i_1}) \cdot f_{i_2}(y_{i_2}) \cdot \ldots \cdot f_{i_n}(y_{i_n}) \quad (y \in X).$$

This is a finite product of continuous real functions (for instance, $y \mapsto f_{i_1}(y_{i_1})$ is $f_{i_1} \circ \pi_{i_1}$, a composite of continuous functions, etc.) and is therefore continuous.
Also, $f(x) = 1 \cdot 1 \cdot \ldots \cdot 1 = 1$
and for any $z \in F \subseteq X \setminus B$ there must be one of the i_j such that $z_{i_j} \notin G_{i_j}$, therefore $f_{i_j}(z_{i_j}) = 0$, therefore $f(z) = 0$.

Hence (X, τ) is $T_{3\frac{1}{2}}$.

For a proof that the $T_{3\frac{1}{2}}$ property is hereditary, please see Essential Exercise 64 and its specimen solution.

6.20 Suppose $(X, \tau) = (X, \tau_d)$, d being a metric.
We already know the space is T_1 (6.2(i)).
Given disjoint non-empty closed $F_0, F_1 \subseteq X$, put

$$G = \{x : d(x, F_0) < d(x, F_1)\}, H = \{x : d(x, F_1) < d(x, F_0)\}.$$

Routine (metric) arguments show G and H open, disjoint, $F_0 \subseteq G, F_1 \subseteq H$.
So (X, τ) is T_4.

6.21 Notice first that if G_0 and G_1 are open sets with $\overline{G_0} \subseteq G_1$, then $\exists$ open $G_{\frac{1}{2}}$ such that $\overline{G_0} \subseteq G_{\frac{1}{2}}$ and $\overline{G_{\frac{1}{2}}} \subseteq G_1$ (see Essential Exercise 62, whose argument is essentially the same as that of 6.15).
Given closed disjoint non-empty F_0 and F_1 in the T_4 space (X, τ), choose open G_0, H_0 to separate them as in 6.19. Set $G_1 = X \setminus F_1$.
Since $\overline{G_0} \cap F_1 = \emptyset$, we have $\overline{G_0} \subseteq G_1$. Then we can construct:

(i) $G_{\frac{1}{2}} \in \tau$ such that $\overline{G_0} \subseteq G_{\frac{1}{2}}, \overline{G_{\frac{1}{2}}} \subseteq G_1$;

(ii) $G_{\frac{1}{4}}$ and $G_{\frac{3}{4}} \in \tau$ such that $\overline{G_0} \subseteq G_{\frac{1}{4}}, \overline{G_{\frac{1}{4}}} \subseteq G_{\frac{1}{2}}, \overline{G_{\frac{1}{2}}} \subseteq G_{\frac{3}{4}}, \overline{G_{\frac{3}{4}}} \subseteq G_1$;

(iii) and so on!

In this fashion, we generate a family of open sets

$$\{G_r : r = \frac{m}{2^n}, n \geq 1, 0 \leq m \leq 2^n\}$$

such that

$$r_1 < r_2 \Rightarrow \overline{G_{r_1}} \subseteq G_{r_2}.$$

The indexing set (of so-called *dyadic rationals*) is dense in $[0, 1]$: that is, if $0 \leq s < t \leq 1$ then there is some dyadic rational $m/2^n$ between s and t.

Now define

$$f(x) = \begin{cases} \inf\{r : x \in G_r\} & \text{if } x \notin F_1, \\ 1 & \text{if } x \in F_1. \end{cases}$$

Certainly $f : X \to [0,1]$ has $f(F_0) = \{0\}$ and $f(F_1) = \{1\}$. To show f continuous, it is enough (compare 5.5 and think *subbase* instead of *base*) to check that

$$\forall \alpha \in (0,1), f^{-1}([0,\alpha)) \text{ and } f^{-1}((\alpha,1]) \text{ are open.}$$

Well, $f(x) < \alpha \Leftrightarrow \exists$ some dyadic rational r such that $f(x) < r < \alpha$; it follows that $f^{-1}([0,\alpha)) = \bigcup\{G_r : r < \alpha\}$, which is, indeed, open! Again,

$$f(x) > \alpha \Leftrightarrow \exists \text{ dyadic } r_1 < r_2 \text{ such that } \alpha < r_1 < r_2 < f(x)\ldots$$

implying $x \notin G_{r_2}$ and therefore $x \notin \overline{G}_{r_1}$. It follows that $f^{-1}((\alpha,1]) = \bigcup_{r_1>\alpha}(X \setminus \overline{G}_{r_1})$, which is also open.

6.22 T_1 is included in both T_4 and $T_{3\frac{1}{2}}$.
Given (X,τ) to be T_4 and $p \notin$ closed F in the space,
F and $\{p\}$ are disjoint closed sets, so Urysohn's lemma tells us that

$$\exists \text{ continuous } f : X \to [0,1] \text{ such that } f(F) = \{0\} \text{ and } f(\{p\}) = \{1\},$$

that is, $f(p) = 1$ and $f(F) = \{0\}$.
So X is $T_{3\frac{1}{2}}$.

6.23 T_1 is implied by T_2 and by normality.
Given disjoint, closed, non-empty F_0, F_1 in a compact T_2 space (X,τ), notice that F_0 and F_1 are compact (4.14),
so 6.8 tells us $\exists$ disjoint open G, H such that

$$F_0 \subseteq G, F_1 \subseteq H.$$

Hence X is T_4.

6.25 Suppose normal (X,τ) has a countable base $\mathcal{B} = \{G_n : n \in \mathbb{N}\}$.
Put $\mathcal{G} = \{(G_n, H_n) \in \mathcal{B} \times \mathcal{B} : \overline{G}_n \subseteq H_n\}$: then $\mathcal{G}$ is countable, so the labelling $\{(G_n, H_n) : n \in \mathbb{N}\}$ is legitimate. For each n, 6.21 allows the choice of continuous $f_n : X \to [0,1]$ such that $f_n(\overline{G}_n) = \{0\}$ and $f_n(y) = 1\ \forall y \in X \setminus H_n$. For each x and y in X, define

$$d(x,y) = \sqrt{\sum_1^\infty \left\{\frac{f_n(x) - f_n(y)}{2^n}\right\}^2}.$$

To confirm that the series converges and that d is a metric is mostly routine. It remains to show that $\tau_d = \tau$.

- Given $x \in G \in \tau$, pick $H \in \mathcal{B}$ such that $x \in H \subseteq G$ and then (see 6.15) find $O_1 \in \tau$ with $x \in O_1 \subseteq \overline{O}_1 \subseteq H$, and $G \in \mathcal{B}$ such that $x \in G \subseteq O_1$. Then (G, H) belongs to the family $\mathcal{G}$, and is therefore (G_n, H_n) for some choice of $n \in \mathbb{N}$.
  Since $(G_n, H_n) \in \mathcal{G}$, we have $f_n(x) = 0$ and $f_n(y) = 1$ $\forall y \in X \setminus H_n$, giving $d(x, y) \geq 2^{-n}$: that is, $B(x, 2^{-n}) \subseteq H_n \subseteq G$.
  So G is a τ_d-neighbourhood of each of its points, therefore τ_d-open. Hence $\tau \subseteq \tau_d$.

- Given $x \in G \in \tau_d$, pick ε such that $B(x, \varepsilon) \subseteq G$, and then pick n such that $2^{-n} < \frac{\varepsilon}{2}$. For $1 \leq j \leq n$, use the continuity of f_j at x to select a τ-neighbourhood V_j of x such that $|f_j(x) - f_j(y)| < \varepsilon/(2n)$ $\forall y \in V_j$.
  Then $V = \bigcap_{j=1}^{n} V_j$ is a τ-neighbourhood of x and, $\forall y \in V$,

$$\begin{aligned}[d(x, y)]^2 &\leq \sum_{j=1}^{n} \left(\frac{f_j(x) - f_j(y)}{2^j} \right)^2 + \sum_{j=n+1}^{\infty} \left(\frac{1}{2^j} \right)^2 \\ &< n\,(\varepsilon/(2n))^2 + \frac{1}{2^{2n+2}} \left(1 + \frac{1}{4} + \frac{1}{16} + \frac{1}{64} + \dots \right) \\ &= \frac{\varepsilon^2}{4n} + \frac{4}{3} \left(\frac{1}{2^{2n+2}} \right) < \frac{\varepsilon^2}{4} + \frac{\varepsilon^2}{12} < \varepsilon^2.\end{aligned}$$

Thus $V \subseteq B(x, \varepsilon) \subseteq G$, so G is a τ-neighbourhood of x for each $x \in G$, and is therefore τ-open; hence $\tau_d \subseteq \tau$.
The proof concludes!

6.27

(iv) Let $x \neq y$ in $(C, \tau(\leq))$. Without loss of generality, we shall take $x < y$.

Case 1: If $\exists z$ such that $x < z < y$
then $(\ \ , z), (z,\ \ )$ are disjoint open sets containing x and y, respectively.

Case 2: If there is *no element* in the interval (x, y)
then $(\ \ , y), (x,\ \ )$ are disjoint . . . and they are open sets containing x and y, respectively.

Hence $\tau(\leq)$ is T_2.

6.28

(i) If W has *no* top element then it is easy to see that

$$\{(\quad, x) : x \in W\}$$

is a family of open intervals that covers W but has no finite subcover. So $(W, \tau(\leq))$ is not compact.

(ii) Now suppose W *has* a top element t, and that C is a family of semi-infinite open intervals that covers W.
Since t never lies in $(\quad, x)$ for $x \in W$, C must include at least one interval of the form $(x, \quad)$.
Put $K = \{x \in W : (x, \quad) \in C\} \neq \emptyset$.
Then K has a least element λ: that is, $(\lambda, \quad)$ does belong to C, but if $\mu < \lambda$ then $(\mu, \quad)$ cannot belong.
Since C covers all of W, including the point λ, some element of C has to include λ *and it must be* of the form $(\quad, \zeta)$.
Then $(\quad, \zeta)$ and $(\lambda, \quad)$ between them cover all of W!
By Alexander's subbase theorem – keeping in mind that the semi-infinite open intervals form a subbase (6.27(iii)) for $\tau(\leq)$ – the space $(W, \tau(\leq))$ is compact.

6.29

(i) • Any neighbourhood N of t contains a basic neighbourhood of the form

$$(u, \quad) = (u, t],$$

where $u < t$. Now $[0, t] = W$ is uncountable, but $[0, u)$ is countable. So certainly $[u, t]$ is uncountable and is *not just* $\{u, t\}$. That shows us that $(u, t]$ contains *many* elements of $(\quad, t)$, and therefore so must N.
By 2.14, $t \in \overline{(\quad, t)}$.

• Let $(x_n)_{n \geq 1}$ be any sequence in $(\quad, t)$.
Because 'every countable set in the first uncountable ordinal is bounded above in it', there is $\lambda < t$ such that

$$x_n \leq \lambda \text{ for every } n \in \mathbb{N}.$$

Then $(\lambda, t]$ is a neighbourhood of t in W including no term of the sequence, so $x_n \not\to t$.

- Now 3.26 tells us that W is not first-countable.

(ii) Take the identity map on $(\quad, t)$ considered as a map into $(\quad, t]$:

$$\text{id} : (\quad, t) \to (\quad, t].$$

Straight off the definitions this is a net of elements of $(\quad, t)$ converging in W to the limit t.

6.30

(i) $W \times Y$ is T_4,
therefore $W \times Y$ is $T_{3\frac{1}{2}}$ (via 6.22).
Therefore all of its subspaces, including TP, must be $T_{3\frac{1}{2}}$ (6.18(iv)).

(ii) In TP, the 'right-hand edge' $\{t\} \times (Y \setminus \{1\})$ equals $TP \cap (\{t\} \times Y)$ and is therefore closed in TP.
Likewise, the 'top edge' $[0, t) \times \{1\}$ is closed *in TP*, and certainly disjoint from $\{t\} \times (Y \setminus \{1\})$.
If TP were T_4 there would have to exist open sets G, H *in* $W \times Y$ such that

$$\{t\} \times (Y \setminus \{1\}) \subseteq G, [0, t) \times \{1\} \subseteq H, G \cap H \cap TP = \emptyset \qquad (*)$$

(Fig. 6.7).

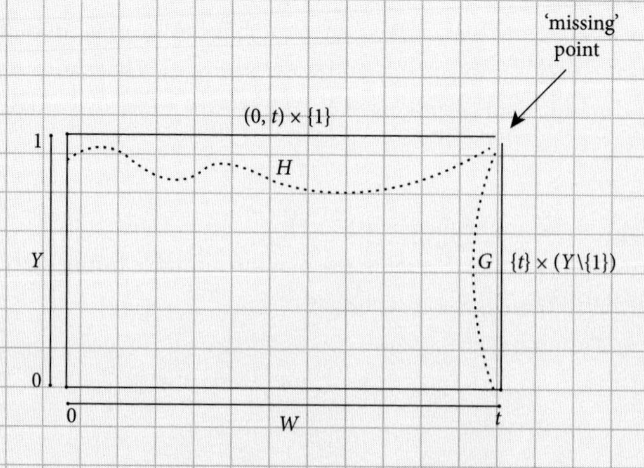

Fig. 6.7 The Tychonoff plank (see Expansion of 6.30).

For each $n \in \mathbb{N}$, the point $\left(t, 1 - \frac{1}{n}\right)$ is inside G, so some open box including $\left(t, 1 - \frac{1}{n}\right)$ lies within G. In particular, $\exists x_n < t$ so that

$$(x_n, t] \times \left\{1 - \frac{1}{n}\right\} \subseteq G.$$

But $\{x_1, x_2, x_3, x_4, \ldots\}$ has a (strict) upper bound $m < t$, and so

$$[m, t] \times \left\{1 - \frac{1}{n}\right\} \subseteq G \text{ for every } n \in \mathbb{N}.$$

We see that the sequence $\left(\left(m, 1 - \frac{1}{n}\right)\right)_{n \geq 1}$ of points of G converges (as $n \to \infty$) to $(m, 1) \in H$, so for all sufficiently large n it has to be true that $\left(m, 1 - \frac{1}{n}\right) \in G \cap H$, contradicting $(*)$ (Fig. 6.8).

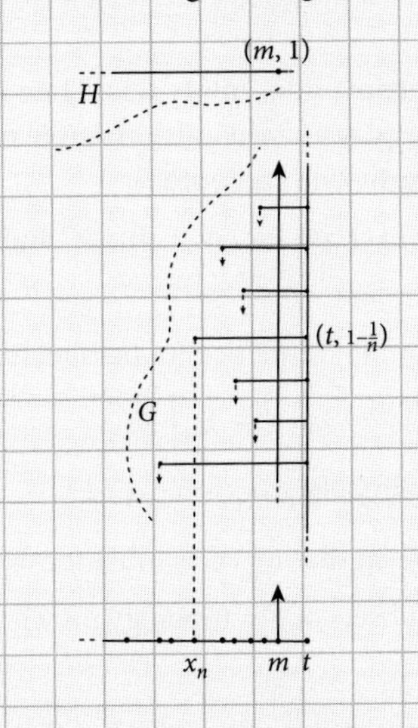

Fig. 6.8 More on the Tychonoff plank (see endgame in the Expansion of 6.30).

Hence the result!

Essential Exercises

1. Given $f : X \to Y$,
   (a) if $f(f^{-1}(B)) = B$ for *every* $B \subseteq Y$, show that f is onto;
   (b) if $f^{-1}(f(A)) = A$ for *every* $A \subseteq X$, show that f is one-to-one.

2. Given sets A, B and C with $A \neq \emptyset$,
   (a) if $A \times B = A \times C$, show that $B = C$;
   (b) if $A \times B = B \times A$, show that either $B = A$ or $B = \emptyset$.

3. (i) Let A, B be given distinct non-empty sets. Then $A \times B$ and $B \times A$ are not, of course, equal. Display an (obvious) example of a mapping from $A \times B$ to $B \times A$ that is one-to-one and onto.
   (ii) Carry out a similar exercise for $(A \times B) \times C$ and $A \times (B \times C)$.

4. Let $L_1, L_2, L_3, \ldots$ be a family (sequence) of parallel lines in the plane, and put
   $$X = \bigcup_{n \geq 1} L_n.$$
   Let us call a subset G of X *big* if *either* $G = \emptyset$ *or*
   $$\exists n_0 \in \mathbb{N} \text{ such that } L_n \setminus G \text{ is finite for every } n \geq n_0.$$
   Is the family of all *big* subsets of X a topology on X?

5. Suppose for the moment that we call a subset of the real line ($\mathbb{R}$) *fat* if it includes all but finitely many of the rational numbers *and* all but countably many of the irrational numbers. Verify that all the fat subsets of $\mathbb{R}$, together with $\emptyset$, comprise a topology on $\mathbb{R}$.

6. Suppose for the moment that we call a subset G of the positive integers ($\mathbb{N}$) *factoid* if:
   $$\text{whenever } x \in G, \text{ then all the factors of } x \text{ belong to } G \text{ also.}$$
   Verify that the collection of factoid subsets is a topology on $\mathbb{N}$.

7. Find or devise examples of spaces in which . . .
   (a) the intersection of any *countable* collection of open sets is open;
   (b) ditto, but some *uncountable* collections of open sets have intersections that are not open;

(c) the closure of every open set is open;

(d) ditto, but some non-empty open set has a closure that is *not* the whole space.

8. Give a (simple) example to show that a point in a topological space may fail to have a smallest neighbourhood: that is, the intersection of *all* neighbourhoods of a point may fail to be a neighbourhood. In contrast, show that in the space of Essential Exercise 6, every point *does* have a smallest neighbourhood.

9. If $f : [a, b] \to \mathbb{R}$ is a continuous real function (in the usual sense) on a closed bounded interval, prove that its graph

$$\Gamma = \{(x, f(x)) : a \leq x \leq b\}$$

is closed in $\mathbb{R}^2$ with the usual topology. (A metric-space argument with sequences is recommended.)

10. For the real function $f : \mathbb{R} \to \mathbb{R}$ given by

$$f(x) = \begin{cases} x \sin\left(\frac{1}{x}\right) & \text{if } x \neq 0, \\ 1 & \text{if } x = 0, \end{cases}$$

determine the closure of its graph.

11. The 'topologist's sine curve' is the graph of the real function $\sin\left(\frac{1}{x}\right)$ $(x \neq 0)$. That is, it is the set

$$\left\{\left(x, \sin\left(\frac{1}{x}\right)\right) : x \in \mathbb{R}, x \neq 0\right\}$$

in the coordinate plane. Determine its closure (in $\mathbb{R}^2$ with its natural metric topology).

12. Find a function $f : \mathbb{R} \to \mathbb{R}$ whose graph is closed but which is not everywhere continuous. (Start with $1/x$, perhaps.)

13. Given a space (X, τ) and $A \subseteq X$, we define its *τ-frontier* $Fr^{\tau}(A)$ thus:

$$x \in Fr^{\tau}(A) \Leftrightarrow \text{every } \tau\text{-neighbourhood of } x \text{ meets both } A \text{ and } X \setminus A.$$

Now, if $B \subseteq A$, show that $Fr^{\tau_A}(B) \subseteq A \cap Fr^{\tau}(B)$. Also, give an (easy) example where equality does not hold.

14. Let $p \in \mathbb{R}$ and let ι_p and ϵ_p be the included-point topology and the excluded-point topology based at p.

    (1) *Which* maps from $(\mathbb{R}, \iota_p)$ to $(\mathbb{R}, \tau_{\text{usual}})$ are continuous?

    (2) *Which* maps from $(\mathbb{R}, \epsilon_p)$ to $(\mathbb{R}, \tau_{\text{usual}})$ are continuous?

15. The interior A° (or $A^{\circ\tau}$) of a subset A in a space (X, τ) is defined to be the union of all the τ-open subsets that are contained in A. In this notation,

    (i) show that $A^\circ = X \setminus \overline{X \setminus A}$;

    (ii) if $A \subseteq B \subseteq X$, show that $A^{\circ\tau_B} \supseteq A^\circ$;

    (iii) show by example that it is possible for $A^{\circ\tau_B} \supset A^\circ$.

16. Given a non-constant map $f : (\mathbb{R}, \tau_{cc}) \to (\mathbb{R}, \tau_{cf})$, show that f is continuous if and only if, for each $y \in \mathbb{R}$, the set $f^{-1}(\{y\})$ is countable.

17. Find (or invent) topological properties that are

    (a) open-hereditary but not closed-hereditary,

    (b) closed-hereditary but not open-hereditary,

    (c) open-hereditary and closed-hereditary but not hereditary (this part is rather difficult!).

    (If you wish, you may address any or all of the parts of this question in the context of metric spaces – where you may well have more experience – instead of in that of topological spaces.)

18. Are the following pairs of spaces homeomorphic?

    (i) $\mathbb{Q} \cup (0, 1)$ and $\mathbb{R} \setminus \mathbb{Q}$ (as subspaces of $(\mathbb{R}, \tau_{\text{usual}})$);

    (ii) $(\mathbb{C}, \tau_{\text{disc}})$ and $(\mathbb{C}, \tau_{cc})$;

    (iii) a circle with one point removed (in $\mathbb{R}^2$ with τ_{usual}) and $(\mathbb{R}, \tau_{\text{usual}})$;

    (iv) $\mathbb{N}$ with topology $\{1, 2\}, \{3, 4\}, \{5, 6\}, \{7, 8\}, \ldots$ and all unions of these, and $\mathbb{N}$ with topology $\{1, 3\}, \{2, 4\}, \{5, 7\}, \{6, 8\}, \{9, 11\}, \{10, 12\}, \ldots$ and all unions of these.

19. Are the following pairs of spaces homeomorphic? Give evidence for your answers.

    (a) $(\mathbb{R}, \tau_{cc})$ and $(\mathbb{R}, \tau_{cf})$;

    (b) $\mathbb{R}^2$ and the surface of a sphere with one point removed (natural metric topologies here);

(c) $(\mathbb{Q}, \iota_x)$ and $(\mathbb{Q}, \iota_y)$, where x and y are two distinct rational numbers;

(d) $[0, 1]$ and $[0, \infty)$ as subspaces of $(\mathbb{R}, \tau_{\text{usual}})$.

20. Suppose that $f : (X, \tau) \to (Y, \sigma)$ is a map, that $X = A \cup B$, and that the two restrictions

$$f|_A : (A, \tau_A) \to (Y, \sigma)$$
$$f|_B : (B, \tau_B) \to (Y, \sigma)$$

are both continuous (on the subspaces). Show that

(i) if A and B are both τ-open, then f is continuous;

(ii) if A and B are both τ-closed, then f is continuous;

(iii) if A is τ-open and B is τ-closed, then it is not necessarily true that f is continuous.

21. If $f : (X, \tau) \to (Y, \sigma)$ is a map, show that it is continuous if and only if: $\forall x \in X$, for every σ-neighbourhood N of $f(x)$, $f^{-1}(N)$ is a τ-neighbourhood of x.

22. On the real interval $[0, 1)$, we define a topology τ by declaring that, for $G \subseteq [0, 1)$, G is τ-open if and only if *either* $0 \notin G$ *or* $[0, 1) \setminus G$ is countable.

(i) Show that (in this space) a sequence $(x_n)_{n \in \mathbb{N}}$ converges to a limit l if and only if $x_n = l$ for all sufficiently large n.

(ii) Decide whether or not $([0, 1), \tau)$ is homeomorphic to $(\mathbb{R}, \tau_{cc})$.

23. Suppose that $(D, \leq')$ and $(E, \leq'')$ are two directed sets where $D \cap E = \emptyset$. We impose an order $\leq$ on $D \cup E$ thus:

$$x \leq y \iff \begin{cases} (x \text{ and } y \in D \text{ and } x \leq' y) \text{ or} \\ (x \text{ and } y \in E \text{ and } x \leq'' y) \text{ or} \\ x \in D \text{ and } y \in E. \end{cases}$$

Check that this makes $D \cup E$ into a directed set.

When will a net $(x_y)_{y \in D \cup E}$ converge in a space? Check that your intuition is right.

24. (i) Given two directed sets $(D, \leq)$ and $(E, \leq')$, suppose that we impose a binary relation $\leq^*$ on the product set $D \times E$ as follows:

$$(d_1, e_1) \leq^* (d_2, e_2) \Leftrightarrow (d_1 \leq d_2 \text{ and } e_1 \leq' e_2).$$

Verify that $D \times E$ is now a directed set.

(ii) Suppose that we also have two convergent nets $(x_\gamma)_{\gamma\in D}, (y_\epsilon)_{\epsilon\in E}$ of real numbers, and we define the new net

$$\left(x_\gamma + y_\epsilon\right)_{(\gamma,\epsilon)\in D\times E}$$

indexed by $(D \times E, \leq^*)$. What do you expect of this net as regards its convergence? Verify or refute your intuition!

25. Suppose that $(x_\gamma)_{\gamma\in D}$ is a convergent net in (X, τ), where $(D, \leq)$ is a directed set. Take any γ_0 in D, define $D_0 = \{\gamma \in D : \gamma \geq \gamma_0\}$ and order the elements of D_0 by the same order as in D (but we'll write it just as $\leq$ rather than as $\leq|_{D_0}$, which is too cumbersome a notation). Show that D_0 is a directed set, and that the net $(x_\gamma)_{\gamma\in D_0}$ also converges.

26. A subset E of a directed set $(D, \leq)$ is said to be *cofinal in D* if, for each $d \in D$, there exists $e \in E$ such that $d \leq e$. (It is easily verified that E is then also a directed set under the same order as D possessed.) If $(x_\gamma)_{\gamma\in D}$ is a net indexed by D, then its restriction to a cofinal subset E, naturally written as $(x_\gamma)_{\gamma\in E}$, is called a *cofinal subnet* of $(x_\gamma)_{\gamma\in D}$. Prove that if a net converges to a limit l then each of its cofinal subnets also converges to the same limit l.

27. (i) If $(x_\gamma)_{\gamma\in D}$ is a net in a *metric* space (M, d), show that it is impossible for it to converge to two *different* limits. (You will need to use the fact that D is directed.)

    (ii) Let f, g be two continuous maps from (X, τ) to a *metric* space (M, d), and put $A = \{x \in X : f(x) = g(x)\}$. Use 3.37 and (i) above to show that A is τ-closed.

28. In Essential Exercise 4, given an arbitrary sequence (z_n) in X, verify that *either* some subsequence of (z_n) lies on one single line L_q *or* there is a subsequence and a non-empty open set G such that no term in the subsequence belongs to G.

29. For each $n \in \mathbb{N}$, let $C_n = \{(n, 1), (n, 2), (n, 3), (n, 4), \ldots\}$, and put $X = \bigcup_{n\geq 1} C_n$ and $X^+ = X \cup \{(0, 0)\}$. Give X^+ a topology by declaring $G \subseteq X^+$ open if and only if:

    either $(0,0) \notin G$

    or $\exists\, n_0 \in \mathbb{N}$ such that $\forall\, n \geq n_0,\ C_n \setminus G$ is finite.

    Prove that no sequence in X can converge to $(0, 0)$.

30. Decide whether or not $(\mathbb{N}, \tau_{cf})$ is sequentially compact.

31. Decide whether the space described in Essential Exercise 6 is
    (i) locally compact,
    (ii) compact,
    (iii) sequentially compact.

32. Decide whether $(\mathbb{Q}, \iota_0)$, $(\mathbb{R}, \epsilon_0)$ are sequentially compact, compact, locally compact.

33. Show that $(\mathbb{R}, \tau_{cc})$ is not compact.

34. Show that (within a given topological space (X, τ)) the union ...
    (i) ... of finitely many compact subsets must be compact,
    (ii) ... of finitely many sequentially compact subsets must be sequentially compact,
    (iii) ... of a countable family of separable subsets must be separable.

35. Consider the following two definitions.
    (a) We call a space *σ-compact* if it can be expressed as the union of a countable family of compact subsets: for instance, the (non-compact) real line $\mathbb{R}$ is σ-compact because $\bigcup_{n \in \mathbb{N}}[-n, n] = \mathbb{R}$. Show that σ-compactness is closed-hereditary.
    (b) We call a space *Lindelöf* if each of its open covers has a *countable* subcover. Show that this property is closed-hereditary but not open-hereditary. (Hint for the second part: observe that an uncountable discrete space is not Lindelöf.)

36. Suppose we are given a space (X, τ) and an open covering $\{G_\alpha : \alpha \in I\}$ that has no finite subcovering. Put $\mathcal{F}(I)$ = the set of all finite subsets of the indexing set I, ordered by set inclusion. Then $\mathcal{F}(I)$ is a directed set – for each $A, B \in \mathcal{F}(I)$, $A \cup B$ is an upper bound in $\mathcal{F}(I)$ for A and B. Also, for each $F \in \mathcal{F}(I)$, $\bigcup_{\alpha \in F} G_\alpha$ is a *proper* subset of X, so we can choose a point $x_F \in X \setminus \bigcup_{\alpha \in F} G_\alpha$.

    Show that the net $(x_F)_{F \in \mathcal{F}(I)}$ has no convergent cofinal subnet (see Essential Exercise 26 for the appropriate definition). *(Remark: this is a step in the direction of an important result, that a space is compact if and only if every net has a convergent subnet; we are avoiding covering this result because the full definition of 'subnet' is rather difficult and, at the level at which we are examining topological ideas, there are usually easier ways to proceed.)*

37. We call (X, τ) *Lindelöf* if every open cover of X has a *countable* subcover. Show that this property is preserved by continuous surjections.

38. Let us, for the moment, call (X, τ) *oligarchical* if $\exists$ finite $A \subseteq X$ such that $\overline{A} = X$. Is this property preserved by continuous surjections?

39. Determine whether oligarchicality (as described in Essential Exercise 38) is hereditary, open-hereditary, closed-hereditary.

40. If A and B are connected subsets of a space (X, τ) and $A \cap B \neq \emptyset$, show that $A \cup B$ is connected.

41. Assuming the result of Essential Exercise 40:
   (i) If $C_1, C_2, \ldots, C_n$ are connected subsets of (X, τ) and
   $$\forall i \in \{1, 2, \ldots, n\}\ \exists j \in \{1, 2, \ldots, n\} \setminus \{i\} \text{ such that } C_i \cap C_j \neq \emptyset, \quad (*)$$
   does it follow that $\bigcup_1^n C_k$ is connected?
   (ii) Does it make a difference if we change $(*)$ to
   $$\forall i \in \{2, 3, \ldots, n\}\ \exists j \in \{1, 2, \ldots, i-1\} \text{ such that } C_i \cap C_j \neq \emptyset?$$

42. Suppose that p is an element in (X, τ) and $\{C_\alpha : \alpha \in I\}$ is an arbitrary family of connected subsets of X, where $p \in C_\alpha$ for every α. Show that $\bigcup_{\alpha \in I} C_\alpha$ is connected.

43. Given a space (X, τ), we define a binary relation $\sim$ on X thus:
   $$x \sim y \text{ iff there is a connected subset of } X \text{ including } x \text{ and } y.$$
   Show that $\sim$ is an equivalence relation.
   Show also that each equivalence class is τ-closed.

44. Suppose that $\mathcal{C}$ and $\mathcal{D}$ are two bases for two different topologies γ and δ on the same set X. Put τ to consist of $\emptyset$ and every possible union of sets of the form $C \cap D$, where $C \in \mathcal{C}$ and $D \in \mathcal{D}$. Is τ a topology on X? (Concentrate on the tricky bit: if, in the obvious notation,
   $$\bigcup_\alpha (C_\alpha \cap D_\alpha) \text{ and } \bigcup_\beta (C_\beta \cap D_\beta)$$
   are not disjoint, is their intersection a union of sets of the form $C' \cap D'$, where $C' \in \mathcal{C}$ and $D' \in \mathcal{D}$?)

45. Suppose we are given a non-empty set X (having no structure yet!) and a family $\mathcal{B}$ of subsets of X. Show that the following are equivalent:

    (a) there is a topology τ on X for which $\mathcal{B}$ is a base;

    (b) $\mathcal{B}$ covers X and, whenever $B_1, B_2 \in \mathcal{B}$ and $x \in B_1 \cap B_2$, there is a $B_3 \in \mathcal{B}$ such that $x \in B_3 \subseteq B_1 \cap B_2$.

46. Show that a base for the discrete topology on an uncountable set cannot be countable.

47. Suppose that you were given a sequence $M_1, M_2, M_3, \ldots$ of neighbourhoods of a point p in a space (X, τ) such that

    $$\text{for every neighbourhood } N \text{ of } p,\ \exists n \in \mathbb{N} \text{ such that } M_n \subseteq N.$$

    How could you modify the given material to produce a countable local base at p?

48. Prove that:

    (i) complete separability implies first-countability;

    (ii) first-countability does not imply complete separability.

49. Show that complete separability is preserved by continuous open surjections.

50. Use the Arens–Fort space to show that complete separability is *not* preserved by continuous surjections.

51. Given an arbitrary cover $\mathcal{S}$ of a non-empty set X, show that there is a topology τ on X for which $\mathcal{S}$ is a subbase. Show also that, if τ' is any topology on X, then $\mathcal{S} \subseteq \tau'$ if and only if $\tau \subseteq \tau'$.

52. Amongst the various possible products of four topological spaces A, B, C and D, show that $(A \times (B \times C)) \times D$ is homeomorphic to $(A \times B) \times (C \times D)$.

53. For a given topological space (X, τ), let (X^2, τ^2) and (X^3, τ^3) denote the products of two and of three copies of (X, τ), respectively. Show that $(X, \tau) \times (X^2, \tau^2)$ is homeomorphic to (X^3, τ^3).

54. Let ζ be an arbitrary topology on the product set $\prod_{i \in I} X_i$ underlying the product space of a family $\{(X_i, \tau_i) : i \in I\}$ of spaces. Show that every projection π_i is (ζ, τ_i)-continuous if and only if the product topology τ is $\subseteq \zeta$.

55. Let (X, τ) be the product of a (possibly infinite) family of spaces $\{(X_i, \tau_i)\}_{i \in I}$. For each of the following invariants 'P,' show that, if X is a P space, then every (X_i, τ_i) must be a P space:

    (a) P = separable,

    (b) P = compact,

    (c) P = locally compact,

    (d) P = completely separable.

56. (i) For a given continuous function $f : (X, \tau) \to (Y, \sigma)$, we define the graph of f (and let us denote it by $\Gamma(f)$, say) in the natural way,

    $$\Gamma(f) = \left\{(x, f(x)) : x \in X\right\},$$

    as a subspace of the product of the two spaces. Show that $\Gamma(f)$ is homeomorphic to X.

    (ii) From this, deduce 5.23.

57. Is it true that $(X, \tau) \times (Y, \sigma)$ is first-countable if and only if both (X, τ) and (Y, σ) are first-countable?

58. Given (X, τ) consider the subset

    $$\Delta = \{(x, x) : x \in X\}$$

    of the product space $(X, \tau) \times (X, \tau)$. Show that (X, τ) is T_2 if and only if Δ is closed (in the product space).

59. Suppose we are given a T_2 space (X, τ) and a family of pairwise disjoint compact subsets $\{C_i\}_{i \in I}$. Must there exist a family of pairwise disjoint open sets $\{G_i\}_{i \in I}$ such that $C_i \subseteq G_i$ for every i?

60. If, on the same set X, τ_1 is a compact topology and τ_2 is a T_2 topology and $\tau_1 \supseteq \tau_2$, show that $\tau_1 = \tau_2$.

61. If (X, τ) is first-countable but not T_2, construct a sequence in X that has (at least) two different limits. Furthermore, demonstrate an example of a space that is first-countable but not T_2!

62. Show that (X, τ) is $T_4 \iff$ it is T_1 *and*

    $$K \text{ is } \tau\text{-closed and } K \subseteq G \in \tau \Rightarrow \exists H \in \tau \text{ such that } K \subseteq H \subseteq \overline{H} \subseteq G. \tag{1}$$

    (An argument similar to the proof of Lemma 6.15 will suffice.)

63. Which of the separation axioms encountered in Chapter 6 does the Arens–Fort space satisfy? (An application of Essential Exercise 62 is recommended.)

64. Show that the $T_{3\frac{1}{2}}$ property is hereditary.

65. Recall that a space is called *Lindelöf* if every open covering of it possesses a *countable* subcovering. Verify that

    (i) a space is compact iff it is Lindelöf *and* every *countable* open covering of it possesses a finite subcovering,

    (ii) a completely separable (that is, second-countable) space must be Lindelöf.

66. (i) Modify the proof of 5.12 to show that a Lindelöf metric space is completely separable. (See, for example, Essential Exercise 65 for the definition of *Lindelöf.*)

    (ii) Deduce that a compact metrisable space is separable.

67. (i) Given a compact metric space (M, d) in which no two points are at a distance greater than 1 from one another, choose (see Essential Exercise 66(ii)) a sequence $(x_n)_{n\geq 1}$ in M whose range is dense in the space. For each $x \in M$ put $\theta(x) = (d(x, x_n))_{n\geq 1}$, and see that θ is a map from (M, d) into the product space $[0, 1]^{\mathbb{N}}$. Show that θ is one-to-one and continuous.

    (ii) Deduce that every compact metrisable space is homeomorphic to a subspace of $[0, 1]^{\mathbb{N}}$.

68. Let (X, τ) be T_3 and Lindelöf, and A and B be non-empty closed disjoint subsets of X.

    (a) Find a countable open cover $\{U_n\}_{n\geq 1}$ of A such that

    $$\overline{U}_n \cap B = \emptyset \ \forall n \geq 1$$

    and a countable open cover $\{V_n\}_{n\geq 1}$ of B such that

    $$\overline{V}_n \cap A = \emptyset \ \forall n \geq 1.$$

    (b) Set $\tilde{U}_n = U_n \setminus (\overline{V}_1 \cup \overline{V}_2 \cup \ldots \cup \overline{V}_n)$, $\tilde{V}_n = V_n \setminus (\overline{U}_1 \cup \overline{U}_2 \cup \ldots \cup \overline{U}_n)$ and

    $$U = \bigcup_{n\geq 1} \tilde{U}_n, \quad V = \bigcup_{n\geq 1} \tilde{V}_n.$$

    Verify that $U \in \tau, V \in \tau, U \cap V = \emptyset, A \subseteq U$ and $B \subseteq V$.

    (c) Deduce that T_3 + Lindelöf $\Rightarrow T_4$.

69. The set $\mathbb{R}^{\mathbb{N}}$ of all sequences of real numbers can be thought of as the product space

$$\mathbb{R} \times \mathbb{R} \times \mathbb{R} \times \mathbb{R} \times \ldots \quad \text{or} \prod_{i=1}^{\infty} \mathbb{R} \quad \text{or} \quad \prod_{i \in \mathbb{N}} \mathbb{R}$$

and given the product topology ($\mathbb{R}$ carrying its usual metric topology).

Suppose now that (X, τ) is a space and $(f_n)_{n \in \mathbb{N}}$ a sequence of continuous functions from X to $\mathbb{R}$ such that

$$\text{for all } x \neq y \text{ in } X, \exists n \in \mathbb{N} \text{ such that } f_n(x) \neq f_n(y).$$

We define $\psi : X \to \mathbb{R}^{\mathbb{N}}$ by

$$\psi(x) = (f_1(x), f_2(x), f_3(x), f_4(x), \ldots).$$

Show that ψ is one-to-one and continuous.

70. Given a $T_{3\frac{1}{2}}$ space (X, τ), suppose that we express the collection of *all* continuous functions from X to $[0, 1]$ as an (indexed) family

$$\{f_\alpha : \alpha \in I\}$$

and, for each α in the (labelling) set I, let Y_α be just the unit interval $[0, 1]$. Now we define a function

$$\theta : X \to \prod_{\alpha \in I} Y_\alpha$$

as follows:

$$\theta(x) = \left(f_\alpha(x)\right)_{\alpha \in I} \qquad (x \in X)$$

(in other words, $\theta(x)$ is the 'vector' whose α^{th} 'component' is $f_\alpha(x)$).

Give reasons why

(i) $\prod_{\alpha \in I} Y_\alpha$ is T_2,

(ii) $\prod_{\alpha \in I} Y_\alpha$ is compact,

(iii) θ is continuous,

(iv) θ is one-to-one.

71. In 6.29, is $W \setminus \{t\}$ first-countable?

72. In 6.29, decide whether the subspace $W \setminus \{t\}$ is locally compact.

73. Change 6.30 by considering the space $W \times W$, the point (t, t) and the 'diagonal'

$$\Delta = \{(x, x) : x \in W \setminus \{t\}\}.$$

Show that $(t, t) \in \overline{\Delta}$ but no sequence in Δ converges to (t, t).

74. In the example in 6.30, decide whether there is

(a) a *sequence* in $(W \times Y) \setminus \{(t, 1)\}$ that converges to $(t, 1)$,

(b) a *sequence* in $(W \times Y) \setminus (\{t\} \times Y)$ that converges to $(t, 1)$.

75. Give $\mathbb{R}$ *not* its usual topology, but the topology ζ that has

$$\{[a, b) : a < b \text{ in } \mathbb{R}\}$$

as a base. Now let *SP* be the product of two copies of $(\mathbb{R}, \zeta)$ – this is often called the Sorgenfrey plane. By considering 'the line $y = -x$', that is, the subset

$$\{(x, -x) : x \in \mathbb{R}\}$$

of *SP*, show that *SP* is not completely separable (that is, second-countable), *but* verify that it is separable.

76. In the space *SP* of Essential Exercise 75, what is the topology on the subspace

$$\{(x, x) : x \in \mathbb{R}\}?$$

77. Show that there exist continuous surjections

(i) from *Cantor*$_3$ onto $[0, 1]$,

(ii) from $(Cantor_3)^{\mathbb{N}}$ onto $[0, 1]^{\mathbb{N}}$,

(iii) from *Cantor*$_3$ onto $[0, 1]^{\mathbb{N}}$.

78. (i) Let K be a non-empty closed subset of a compact set C in the real line. It is an elementary result in metric space theory (not only in $\mathbb{R}$) that for each $x \in C$ there is a closest element in K to x, that is, $\exists\, k(x)$ in K such that $|x - k(x)| \leq |x - y|$ for every $y \in K$. Supposing that K and C are so arranged that this $k(x)$ is unique for each given x, prove that the map $k : C \to K$ is continuous and onto.

(ii) Deduce that for any given non-empty closed subset K of *Cantor*$_5$ there is a continuous surjection from *Cantor*$_5$ onto K.

79. Show that every compact metrisable space is a continuous image of the Cantor space (that is, given an arbitrary compact metrisable space (X, τ), there is a continuous surjection $\gamma : Cantor \to (X, \tau)$).

Solutions to selected exercises

3. (i) A typical element of $A \times B$ takes the form

$$(x, y) \quad \text{where } x \in A \text{ and } y \in B.$$

We notice that (y, x) is then an element of $B \times A$.

So, define

$$f : A \times B \to B \times A$$

by the rule

$$f((x, y)) = (y, x) \quad \text{whenever } (x, y) \in A \times B.$$

It is pretty easy to check that f is one-to-one and onto. A quick and convincing way to confirm this is to define another map

$$g : B \times A \to A \times B$$

by the rule

$$g((y, x)) = (x, y) \quad \text{whenever } (y, x) \in B \times A.$$

Notice that f and g are inverses of one another, and recall that *only* bijections have inverses.

(ii) A typical element of $(A \times B) \times C$ takes the form

$$((x, y), z) \quad \text{where } x \in A, y \in B \text{ and } z \in C.$$

Define

$$h : (A \times B) \times C \to A \times (B \times C)$$

by

$$h(((x, y), z)) = (x, (y, z)) \quad (x \in A, y \in B, z \in C)$$

and, likewise, $j : A \times (B \times C) \to (A \times B) \times C$ by

$$j((x, (y, z))) = ((x, y), z).$$

Once again h and j cancel one another out, so they are mutually inverse, and therefore h is one-to-one and onto.

5. Clearly $\mathbb{R}$ itself is 'fat', and $\emptyset$ is written into the definition.

Let $\{G_\alpha : \alpha \in I\}$ be any family of 'fat or empty' sets, and put $G = \bigcup_{\alpha \in I} G_\alpha$.

If all the G_α are empty, so is G.

If not, pick a particular $\alpha_0 \in I$ so that G_{α_0} is fat.

Then $\mathbb{Q} \setminus G_{\alpha_0}$ is finite and $(\mathbb{R} \setminus \mathbb{Q}) \setminus G_{\alpha_0}$ is countable.

But since $G_{\alpha_0} \subseteq G$,

$$\mathbb{Q} \setminus G \subseteq \mathbb{Q} \setminus G_{\alpha_0} \text{ and is also finite,}$$
$$(\mathbb{R} \setminus \mathbb{Q}) \setminus G \subseteq (\mathbb{R} \setminus \mathbb{Q}) \setminus G_{\alpha_0} \text{ and is also countable.}$$

In both cases, G is fat or empty.

Let G_1, G_2 be fat or empty.

If either is empty, then so is $G_1 \cap G_2$.

If not, then
$\mathbb{Q} \setminus (G_1 \cap G_2) = (\mathbb{Q} \setminus G_1) \cup (\mathbb{Q} \setminus G_2)$ is the union of two finite sets, therefore finite, and

$(\mathbb{R} \setminus \mathbb{Q}) \setminus (G_1 \cap G_2) = ((\mathbb{R} \setminus \mathbb{Q}) \setminus G_1) \cup ((\mathbb{R} \setminus \mathbb{Q}) \setminus G_2)$ is the union of two countable sets, therefore countable.

In both cases $G_1 \cap G_2$ is fat or empty. Now a routine induction shows that the intersection of any finite list of fat-or-empty sets is another such set.

6. Clearly $\emptyset$ and $\mathbb{N}$ itself are factoid.

   Let $\{G_\alpha : \alpha \in I\}$ be any family of factoid subsets of $\mathbb{N}$, and put $G = \bigcup_{\alpha \in I} G_\alpha$.

   Now suppose that $x \in G$.
   Then we can pick α' so that $x \in G_{\alpha'}$.
   Since $G_{\alpha'}$ is factoid, all the factors of x belong to $G_{\alpha'}$ and therefore lie within G.

   So G is factoid.

   Continuing, put $H = \bigcap_{\alpha \in I} G_\alpha$.
   Now suppose that $x \in H$.
   For every $\alpha \in I$, $x \in$ factoid G_α, so all factors of x lie in G_α
   and hence all factors of x lie in H.
   So H is factoid.
   (We only needed to do the last paragraph for *finite* intersections but, unusually, in this example the argument goes through for *arbitrary* intersections.)

7. (a) Any discrete space will serve as a (lazy) example since, in such a space, *every* subset is open.

   (b) For instance, in $(\mathbb{R}, \tau_{cc})$, the closed sets are $\mathbb{R}$ itself and the countable sets, so any countable union of closed sets is closed. Use De Morgan's laws on that fact, and we see that any countable intersection of open sets is open.

On the other hand, $\mathbb{R}\setminus\{x\}$ is τ_{CC}-open for each $x \in [0, 1]$, yet $\bigcap_{x\in[0,1]}(\mathbb{R}\setminus\{x\}) = \mathbb{R}\setminus[0, 1]$ is not τ_{CC}-open.

(c) Any trivial space provides an easy example here.

(d) For a simple (but not very obvious) example, consider the space (X, τ) described by $X = \{a, b, c, d\}$, $\tau = \{\emptyset, \{a, b\}, \{c, d\}, X\}$. Its closed sets are exactly the same as its open sets, so the closure of every open set is open. Yet, for instance, $\overline{\{a, b\}}$ is not the whole of X.

8. For instance, consider neighbourhoods of 0 in the real line with its usual metric topology.

Method 1: Suppose there were a *smallest* neighbourhood N. We can pick $\varepsilon > 0$ so that

$$(-\varepsilon, \varepsilon) \subseteq N.$$

But then $(-\frac{1}{2}\varepsilon, \frac{1}{2}\varepsilon)$ is also a neighbourhood of 0 and it is *strictly smaller* than N: contradiction.

Method 2: Put K = the intersection of *all* neighbourhoods of 0.

Certainly $\{0\} \subseteq K$.

But for any $x \neq 0$, $|x| > 0$ and $(-|x|, |x|)$ is neighbourhood of 0 to which x does not belong . . . so x cannot belong to K.

Thus K is just $\{0\}$, and this is not a neighbourhood of 0, since *no* choice of ε will make

$$(-\varepsilon, \varepsilon) \subseteq \{0\}.$$

In Essential Exercise 6, for any $x \in \mathbb{N}$ let F_x denote the set of all factors of x.

If $y \in F_x$ and z is a factor of y, then z is a factor of a factor of x and therefore z is a factor of x; that is, $z \in F_x$.

The last sentence tells us that F_x is factoid, that is, is an open set in this particular topology.

So F_x is a neighbourhood of x.

For any neighbourhood H of x, pick factoid G such that $x \in G \subseteq H$.

Then all factors of x lie in G, so $F_x \subseteq H$.

Thus F_x is the smallest neighbourhood of x in this context.

11. A careful sketch will probably be enough to convince you that no points (x, y) *with* $x \neq 0$ are in the closure of this graph except those that are on the graph! We will look more carefully at this later.

For any point $(0, t)$, where $-1 \leq t \leq +1$, choose a positive 'angle' θ in radians such that $\sin\theta = t$. Then also

$$t = \sin(2\pi + \theta) = \sin(4\pi + \theta) = \sin(6\pi + \theta) = \ldots$$

For any (metric) neighbourhood N of $(0, t)$, choose $\varepsilon > 0$ so that $B((0, t), \varepsilon) \subseteq N$. Next, choose a positive integer n greater then $1/2\pi\varepsilon$. Look:

$$n > \frac{1}{2\pi\varepsilon}, \quad \text{therefore } 2\pi n > \frac{1}{\varepsilon}, \quad \text{therefore } \frac{1}{2\pi n} < \varepsilon,$$

therefore the distance from $(0, t)$ to $\left(\frac{1}{2n\pi + \theta}, \sin\left(\frac{1}{\left(\frac{1}{2n\pi+\theta}\right)}\right)\right)$

$$= \left(\frac{1}{2n\pi + \theta}, \sin(2n\pi + \theta)\right) = \left(\frac{1}{2n\pi + \theta}, t\right)$$

is

$$\frac{1}{2n\pi + \theta} < \frac{1}{2n\pi} < \varepsilon.$$

In other words, we have found a point on the graph inside $B((0, t), \varepsilon)$ and therefore inside N. So N meets the graph. Now 2.14 tells us that $(0, t)$ lies in the closure of the graph.

(If you are happy to use sequences in this metric-space argument then – good! – do so; they are great in that context.)

We now have the segment S that joins the points $(0, -1)$ and $(0, 1)$ lying within the closure of the graph. It is easy to see that *no other* points on the y-axis can do so.

Thus: $\overline{graph} = graph \cup S$.

15. (i) $x \in A^\circ \Leftrightarrow \exists G \in \tau$ such that $x \in G \subseteq A$
$\Leftrightarrow A$ is a neighbourhood of x
$\Leftrightarrow A$ contains some neighbourhood of x
$\Leftrightarrow$ some neighbourhood of x does not intersect $X \setminus A$
$\Leftrightarrow x \notin \overline{X \setminus A}$
$\Leftrightarrow x \in X \setminus \overline{X \setminus A}$.

(ii) If $x \in A^\circ$ then, for some $G \in \tau, x \in G \subseteq A$,
therefore $x \in B \cap G \subseteq B \cap A = A$ (but $B \cap G$ is τ_B-open),
therefore A is a τ_B-neighbourhood of x,
therefore $x \in A^{\circ\tau_B}$.

(iii) Take the case where (X, τ) is the real line (with the usual topology), $B = [0, 2], A = (1, 2]$.
Then $A^\circ = (1, 2)$
but A is open in the subspace B, so $A^{\circ\tau_B} = (1, 2]$.

16. • Suppose f is continuous.

For each $y \in \mathbb{R}$, $\{y\}$ is τ_{cf}-closed,
so $f^{-1}(\{y\})$ must be τ_{cc}-closed ($\neq \mathbb{R}$ since f is non-constant!),
therefore $f^{-1}(\{y\})$ is countable.

• Suppose $f^{-1}(\{y\})$ is countable for each $y \in \mathbb{R}$.

For any closed K in $(\mathbb{R}, \tau_{cf})$, K is either $\mathbb{R}$ or finite.
Now $f^{-1}(\mathbb{R}) = \mathbb{R}$ is τ_{cc}-closed,
and if K is finite, then
$f^{-1}(K) = \bigcup_{y \in K} f^{-1}(\{y\})$ is a finite union of countable sets,
therefore countable, therefore τ_{cc}-closed again.
By 3.2, f is continuous.

17. (a) The property 'every non-empty open set is infinite' is satisfied by $(\mathbb{R}, \tau_{usual})$ but not by its closed subspace $\{1, 2, 3\}$. So it is *not* closed-hereditary.

Yet if (X, τ) possesses this property and (G, τ_G) is a subspace such that $G \in \tau$, then any non-empty τ_G-open set is already τ-open, and consequently infinite. So (G, τ_G) has 'the property'. So 'the property' is open-hereditary.

(b) This may be a good place for switching to metric spaces! Compactness is closed-hereditary here – as is well known. But (with the usual metric) $[0, 1]$ is compact and its 'open' subspace $(0, 1)$ is not, so compactness is not open-hereditary among metric spaces.
(The same turns out to be true in topological spaces.)

(c) Tricky! Consider this:
For the moment, let us call a space (X, τ) 'lumpy' if every non-empty subset that can be expressed as the intersection of a τ-open set and a τ-closed set is infinite.

Suppose that (G, τ_G) is a subspace of a lumpy space (X, τ), where G is τ-open. Next, suppose that $H \cap F$ is a non-empty subset of G in which H is τ_G-open and F is τ_G-closed. We can express H and F in the forms $H = G \cap H'$, $F = G \cap F'$, where H' is τ-open and F' is τ-closed. So $H \cap F = G \cap H' \cap G \cap F' = (G \cap H') \cap F'$, where $(G \cap H')$ is τ-open and F' is τ-closed in X. Since X is a lumpy space, $H \cap F$ is infinite. This shows lumpiness to be open-hereditary. A very similar argument shows it to be closed-hereditary also.

Yet, on the real interval $[0, 2]$, declare the topology σ to comprise only the three subsets $\emptyset$, $[0, 1)$, $[0, 2]$. It is the work of a moment to check that this gives a lumpy space. However, its (non-open, non-closed) subspace $\{1\}$ is trivially not lumpy. So lumpiness is not hereditary in general.

19. (a) No. The statement 'there is a countably infinite closed subset' is (obviously) a homeomorphic invariant, is true in $(\mathbb{R}, \tau_{cc})$ and is false in $(\mathbb{R}, \tau_{cf})$.

    (b) Yes. Imagine that we place S, a spherical surface (in $\mathbb{R}^3$), onto the x–y plane so that its south pole touches it at the origin, and we label its north pole N. For each point x on $S\setminus\{N\}$, the straight line from N to x hits the x–y plane at a point $\theta(x)$. The correspondence between x and $\theta(x)$ is one-to-one and onto and, as x varies, $\theta(x)$ varies continuously and vice versa. So $S\setminus\{N\}$ and $\mathbb{R}^2$ are homeomorphic.

    (c) Yes. The map $h : \mathbb{Q} \to \mathbb{Q}$ given by

    $$h(x) = y,$$
    $$h(y) = x,$$
    $$h(z) = z \quad \text{when } z \neq x \text{ or } y$$

    is routinely checkable to be a homeomorphism.

    (d) No. For instance, one is compact and one is not. Or, one has *only one point* whose complement cannot be split into two open non-empty subsets, and the other has *two such points*. We discuss properties like this more fully in Chapter 4.

21. (a) Suppose that f is continuous.

    Let $x \in X$ and let a σ-neighbourhood N of $f(x)$ be given.

    Choose σ-open G such that $f(x) \in G \subseteq N$.

    Then $x \in f^{-1}(G) \subseteq f^{-1}(N)$, where $f^{-1}(G)$ is τ-open

    and thus $f^{-1}(N)$ is a τ-neighbourhood of x.

    So 'the condition' holds.

    (b) Suppose that 'the condition' holds.

    Let G be any σ-open set (within Y).

    Consider arbitrary $x \in f^{-1}(G)$.

    We see that $f(x) \in G \in \sigma$, so G is a σ-neighbourhood of $f(x)$.

    By 'the condition', $f^{-1}(G)$ is a τ-neighbourhood of x.

    That is, $f^{-1}(G)$ is a τ-neighbourhood of every one of its own points and is therefore τ-open.

So f is continuous.

24. (i)
- For all $d \in D$ and $e \in E$, $d \leq d$ and $e \leq' e$, and therefore

$$(d, e) \leq^* (d, e) \quad (\textit{reflexive!}).$$

- If $(d_1, e_1) \leq^* (d_2, e_2)$ and $(d_2, e_2) \leq^* (d_3, e_3)$, then

$$d_1 \leq d_2, d_2 \leq d_3, e_1 \leq' e_2, e_2 \leq' e_3;$$

therefore $d_1 \leq d_3$ and $e_1 \leq' e_3$, that is, $(d_1, e_1) \leq^* (d_3, e_3)$ (*transitive!*).

- Given any $(d_1, e_1), (d_2, e_2)$ in $D \times E$, *in* D, d_1 and d_2 have a common upper bound, d_3 say;

$$\text{that is, } d_1 \leq d_3 \text{ and } d_2 \leq d_3.$$

Likewise, $\exists e_3 \in E$ such that $e_1 \leq' e_3$ and $e_2 \leq' e_3$.

Then both $(d_1, e_1) \leq^* (d_3, e_3)$ and $(d_2, e_2) \leq^* (d_3, e_3)$ (*directedness!*).

(ii) If $x_\gamma \to l \in \mathbb{R}$ and $y_\epsilon \to m \in \mathbb{R}$, the known behaviour of sequences leads us to expect $x_\gamma + y_\epsilon \to l + m$.

Given $\varepsilon > 0$:

$$\exists \gamma_0 \in D \text{ such that } \gamma \geq \gamma_0 \Rightarrow x_\gamma \in \left(l - \frac{\varepsilon}{2}, l + \frac{\varepsilon}{2}\right)$$

and

$$\exists \epsilon_0 \in E \text{ such that } \epsilon \geq \epsilon_0 \Rightarrow y_\epsilon \in \left(m - \frac{\varepsilon}{2}, m + \frac{\varepsilon}{2}\right).$$

So $(\gamma_0, \epsilon_0) \in D \times E$, and

$$(\gamma, \epsilon) \geq^* (\gamma_0, \epsilon_0) \Rightarrow \gamma \geq \gamma_0 \text{ and } \epsilon \geq' \epsilon_0 \Rightarrow$$

$$|x_\gamma - l| < \frac{\varepsilon}{2} \text{ and } |y_\epsilon - m| < \frac{\varepsilon}{2}$$

$$\Rightarrow \left|(x_\gamma + y_\epsilon) - (l + m)\right| < \varepsilon.$$

So we guessed right.

26. Let $(x_\gamma)_{\gamma \in D}$ converge to l.
    Let $E \subseteq D$ be cofinal, so $(x_\gamma)_{\gamma \in E}$ is a cofinal subnet.
    Let N be any neighbourhood of l.
    We know $\exists \gamma_0 \in D$ such that $\gamma \geq \gamma_0$ $(\gamma \in D) \Rightarrow x_\gamma \in N$.
    Also, $\exists e_0 \in E$ such that $e_0 \geq \gamma_0$.
    Then, for any $e \in E$ such that $e \geq e_0$,
        we know that $e \in D$ and $e \geq \gamma_0$,
        therefore $x_e \in N$.
    Hence $(x_\gamma)_{\gamma \in E}$ converges to l also.

27. (i) *Suppose* if possible that net $(x_\gamma)_{\gamma \in D} \rightarrow$ both to l_1 and to l_2 $(l_1 \neq l_2)$ in the metric space (M, d).
    Put $\varepsilon = \frac{1}{2}d(l_1, l_2) > 0$.
    Then

$$\exists \gamma_1 \in D \text{ such that } \gamma \geq \gamma_1 \Rightarrow x_\gamma \in B(l_1, \varepsilon),$$

$$\exists \gamma_2 \in D \text{ such that } \gamma \geq \gamma_2 \Rightarrow x_\gamma \in B(l_2, \varepsilon).$$

    Since D is *directed*, choose $\gamma_3 \geq$ both γ_1 and γ_2: so

$$x_{\gamma_3} \in B(l_1, \varepsilon) \cap B(l_2, \varepsilon).$$

    By the triangle inequality,

$$d(l_1, l_2) \leq d(l_1, x_{\gamma_3}) + d(x_{\gamma_3}, l_2),$$

    that is,

$$2\varepsilon < \varepsilon + \varepsilon, \quad \text{i.e. } 2\varepsilon < 2\varepsilon, \text{ contradiction!}$$

    (ii) Let $p \in \overline{A}$.
    By 3.37, there is a net $(a_\gamma)_{\gamma \in D}$ of points of A that converges to p.
    Since f is continuous,

$$f(a_\gamma) \rightarrow f(p).$$

    Since g is continuous,

$$g(a_\gamma) \rightarrow g(p).$$

    But $(f(a_\gamma))_{\gamma \in D}, (g(a_\gamma))_{\gamma \in D}$ are the same net! (Since, inside $A, f = g$ always.)
    By (i) we now get $f(p) = g(p)$, that is, $p \in A$. Now $\overline{A} \subseteq A$ is established, and $A = \overline{A}$, and A is closed.

29. (Why is this a topology? Really the same argument as in Essential Exercise 4.)

    Suppose if possible that there is a sequence $(z_i) \to (0, 0)$ in X^+ where each $z_i = (x_i, y_i) \in X$. For the usual reasons, *every* subsequence of (z_i) would have to converge to $(0, 0)$ also.

    Case 1: One individual 'column' C_n might include z_i for infinitely many i. If so, *these* z_i would form a subsequence. And this subsequence fails to converge to $(0, 0)$ because

    $$\{(0, 0)\} \cup C_{n+1} \cup C_{n+2} \cup C_{n+3} \cup \ldots$$

    is an open neighbourhood of $(0, 0)$ which the subsequence fails to enter. Contradiction!

    Case 2: Otherwise, put $N = X^+ \setminus$ (the range of the sequence (z_i)), and this set misses only a finite number of terms in each column, so it is an open neighbourhood of $(0, 0)$ which the sequence fails to enter. Contradiction again!

    Note This is an important example, sometimes called *the Arens–Fort space.* We shall use it.

30. First, here is a noteworthy thing about $(\mathbb{N}, \tau_{cf})$: a sequence with no repeats ends up inside *any* given non-empty open set G! Because $\mathbb{N} \setminus G$ is finite, if you move along such a sequence $(x_n)_{n \geq 1}$ past each n-value for which x_n belongs to $\mathbb{N} \setminus G$, all the later n-values must give $x_n \in G$.

    It follows that a repeat-free sequence converges to every point!

    Now, given *any* sequence (y_n) in $(\mathbb{N}, \tau_{cf}) \ldots$

    (a) if some number turns up infinitely often in the sequence, its occurrences form a subsequence that is constant, and therefore convergent;

    (b) if not, we can lift out a subsequence with no repeats, which also converges.

    So $(\mathbb{N}, \tau_{cf})$ is sequentially compact.

31. (i) For each $n \in \mathbb{N}$, let F_n denote the set of all factors of n (including 1 and n).

    (As we checked in Essential Exercise 8) F_n is factoid, and therefore an open neighbourhood of n. Also, F_n is finite, and therefore certainly compact. So this space is locally compact.

    (ii) $\{F_n : n \in \mathbb{N}\}$ is an open cover of $\mathbb{N}$.

But finitely many of these finite sets can only have a *finite* union; so there is no finite subcover of $\mathbb{N}$.

Thus, the space is not compact.

(iii) Consider the sequence $(n)_{n\in\mathbb{N}}$. All of its subsequences are unbounded.

But if a sequence $(x_n)_{n\in\mathbb{N}}$ converges to l in this space, we must have $x_n \in F_l$ for all sufficiently large n, which forces (x_n) to be bounded.

So no subsequence of $(n)_{n\in\mathbb{N}}$ converges, and the space is not sequentially compact.

33. For instance, take $G_n = (\mathbb{R} \setminus \mathbb{N}) \cup \{n\}$ for each $n \in \mathbb{N}$. Then G_n has countable complement $\mathbb{N} \setminus \{n\}$ and is τ_{cc}-open. Each non-(positive integer) lies in every G_n, and each positive integer lies in exactly one, so

$$\bigcup_{n\in\mathbb{N}} G_n = \mathbb{R}$$

and we have an open cover.

Yet there cannot be any finite subcover: finitely many of the G_n's would capture only finitely many positive integers.

Thus $(\mathbb{R}, \tau_{cc})$ is *not* compact.

34. (i) It will be enough to deal with two subsets (and then use induction).

Let A and B be compact subsets in (X, τ).

Let $\{G_i : i \in I\}$ be any τ-open cover of $A \cup B$.

Then $\{G_i : i \in I\}$ is an open cover of compact A

so $\exists i_1, i_2, \ldots, i_n$ in I such that $A \subseteq \bigcup_{j=1}^{n} G_{i_j}$.

Likewise, $\exists i_{n+1}, i_{n+2}, \ldots i_m$ in I such that $B \subseteq \bigcup_{j=n+1}^{m} G_{i_j}$.

Thus $A \cup B \subseteq \bigcup_{j=1}^{m} G_{i_j}$.

By 4.13, $A \cup B$ is compact.

(ii) Let A and B be sequentially compact subsets in (X, τ).

Let $(x_n)_{n\in\mathbb{N}}$ be any sequence in $A \cup B$.

- If x_n lies in A for infinitely many values of n,

  their occurrences form a subsequence of (x_n) in sequentially compact A,

  so some sub-subsequence converges to a point of A.
- If not, then this argument will work in B instead.

In both cases, some subsequence of (x_n) will have to converge to a limit in $A \cup B$.

So $A \cup B$ is sequentially compact.

Induction extends this to any finite number of sequentially compact subsets.

(iii) Suppose we are given a family $\{A_n : n \in \mathbb{N}\}$ of (countably many) separable subsets of (X, τ). For each $n \in \mathbb{N}$, choose countable $D_n \subseteq A_n$ such that $\overline{D_n}^{\tau_{A_n}} = A_n$,

that is (see 2.18), $\overline{D_n}^{\tau} \supseteq A_n$.

Put $D = \bigcup_{n \in \mathbb{N}} D_n$ and $A = \bigcup_{n \in \mathbb{N}} A_n$.

Certainly D is a countable subset of A.

For each $n \in \mathbb{N}$, $D_n \subseteq D$, therefore $\overline{D}_n \subseteq \overline{D}$,

therefore $A_n \subseteq \overline{D}$,

therefore $A \subseteq \overline{D}$.

Equivalently (2.18 again) $\overline{D}^{\tau_A} = A$, so D is dense in (A, τ_A).

Hence A is separable.

35. (a) Let (X, τ) be σ-compact and A be non-empty and closed in X.

Express X as $\bigcup_{n \in \mathbb{N}} K_n$, where each K_n is compact.

Then

$$A = A \cap X = A \cap \left(\bigcup_{n \in \mathbb{N}} K_n \right)$$
$$= \bigcup_{n \in \mathbb{N}} (A \cap K_n),$$

where each $A \cap K_n$, being closed in the compact subspace K_n, is compact.

Hence A is σ-compact.

(b) The argument of 4.13 shows that a subset A of a space (X, τ) is Lindelöf if and only if every τ-*open* cover of A has a countable subcover: that is, we do not need to work with the subspace topology.

Given A non-empty and closed in a Lindelöf space (X, τ),

let $\{G_j : j \in J\}$ be any τ-open cover of A.

Then $\{X \setminus A, G_j : j \in J\}$ is a τ-open cover of X,

so there is a countable subcover $\{X \setminus A, G_{j_n} : n \in \mathbb{N}\}$.

Therefore $A \subseteq \bigcup_{n \in \mathbb{N}} G_{j_n}$,

that is, $\{G_{j_n} : n \in \mathbb{N}\}$ is a countable subcover of A.

Hence A is Lindelöf.

Consider $(\mathbb{R}, \epsilon_0)$.

Any open cover of $\mathbb{R}$ includes an open set to which 0 belongs

and $\mathbb{R}$ itself is the only such set,

so the open cover has $\{\mathbb{R}\}$ as a *one-set* subcover.

Hence $(\mathbb{R}, \epsilon_0)$ is Lindelöf.

But for each $x \neq 0$ in $\mathbb{R}$, $\{x\}$ is ϵ_0-open,

therefore $\{x\} = (\mathbb{R} \setminus \{0\}) \cap \{x\}$ is $\epsilon_0 |_{\mathbb{R} \setminus \{0\}}$-open;

thus $\mathbb{R} \setminus \{0\}$ as a subspace is discrete and uncountable:

therefore it is *not* Lindelöf.

Also, $\mathbb{R} \setminus \{0\}$ is ϵ_0-open,

so Lindelöfness is not open-hereditary.

36. Suppose, if possible, that $\mathcal{G}$ were a cofinal subset of $\mathcal{F}(I)$ such that the cofinal subnet $(x_F)_{F \in \mathcal{G}}$ converged to some $l \in X$.

Then, for some $\alpha_0 \in I, l \in G_{\alpha_0}$;

thus, G_{α_0} is an open neighbourhood of l.

Convergence tells us $\exists F_0 \in \mathcal{F}(I)$ such that

$$F \supseteq F_0 \Rightarrow x_F \in G_{\alpha_0}. \tag{1}$$

Directedness tells us $\exists F_1 \in \mathcal{F}(I)$ such that $F_1 \supseteq F_0$ and $F_1 \supseteq \{\alpha_0\}$.

Then, by (1), $x_{F_1} \in G_{\alpha_0}$

and yet, by how the x_F's were chosen, $x_{F_1} \notin \bigcup_{\alpha \in F_1} G_\alpha \supseteq G_{\alpha_0}$
– contradiction!

40. Suppose that $A \cup B$ is *not* connected.

By 4.29, $\exists$ τ-open G, H such that

$$A \cup B \subseteq G \cup H, (A \cup B) \cap G \neq \emptyset, (A \cup B) \cap H \neq \emptyset, (A \cup B) \cap G \cap H = \emptyset.$$

It follows that

$$A \subseteq G \cup H \quad \text{and} \quad A \cap G \cap H = \emptyset.$$

So *if* G and H *both* intersected A, we would get a contradiction (because 4.29 would tell us that A was *not* connected).

So A meets *only one* of G, H.

Likewise, B meets *only one* of G, H . . . and it has to be *the other one*, since G and H both meet $A \cup B$.

Without loss of generality, A meets G but not H, and B meets H but not G.

Since $A \cup B \subseteq G \cup H$, this forces $A \subseteq G, B \subseteq H$.

Yet now $A \cap B \subseteq (A \cup B) \cap G \cap H = \emptyset$, forcing $A \cap B = \emptyset$, a contradiction.

43. For each $x \in X$, $\{x\}$ is connected (of course it is!) and includes x and x, so $x \sim x$.

    If connected C includes x and y, then it includes y and x, so $x \sim y \Rightarrow y \sim x$.

    If $x \sim y$ and $y \sim z$, choose connected C_1, C_2 such that x and $y \in C_1$, y and $z \in C_2$.

    Then $C_1 \cup C_2$ is connected (see Essential Exercise 40) and includes x and $z \ldots$ so $x \sim z$.

    For each point y in the equivalence class C_x of x, we know that $x \sim y$ and therefore there is a connected set D_y that includes both x and y. Each element $z \in D_y$ satisfies $z \sim x$ (by virtue of the same connected set D_y), so $D_y \subseteq C_x$. It follows that $\bigcup_{y \in C_x} D_y \subseteq C_x$ and, since the reverse inclusion is immediate, we get the equality

$$\bigcup_{y \in C_x} D_y = C_x.$$

Thus C_x is a union of connected sets all of which include x, and is therefore connected (see Essential Exercise 42 this time). Then $\overline{C_x}$ is connected (by 4.38) and includes x, so every element of $\overline{C_x}$ is related by $\sim$ to x. That is, $\overline{C_x} \subseteq C_x$. It follows that $\overline{C_x} = C_x$, that is, C_x is closed.

45. Suppose (a).
    Then X itself is a non-empty open set, and therefore can be expressed as a union of certain members of $\mathcal{B}$. So $\mathcal{B}$ covers X.

    Now suppose $x \in B_1 \cap B_2$, where $B_1, B_2 \in \mathcal{B}$. By (a), B_1 and $B_2 \in \tau$ and $B_1 \cap B_2 \in \tau$ and $B_1 \cap B_2 \neq \emptyset$: so $B_1 \cap B_2$ is the union of certain elements of $\mathcal{B}$. One of them – call it B_3 – has to include x: so $x \in B_3 \subseteq B_1 \cap B_2$.

    Now (b) is established.

    Suppose (b).

    Define τ to be all unions of subfamilies of $\mathcal{B}$ (and $\emptyset$, of course). Is τ a topology?

    - $\emptyset \in \tau$, and $X \in \tau$ since $\mathcal{B}$ covers X.
    - Any union of members of τ (politely ignoring $\emptyset$ since it does not contribute!) is a union of unions of members of $\mathcal{B}$, therefore still a union of members of $\mathcal{B}$, therefore is still in τ.

- For each x in the intersection of *two* members G, H of τ . . .

$$\ldots \text{ say, } G = \bigcup_{\alpha \in A} B_\alpha, H = \bigcup_{\beta \in I} B_\beta \ldots$$

there must be

$$\alpha' \in A, \beta' \in I \text{ such that } x \in B_{\alpha'} \text{ and } x \in B_{\beta'}.$$

Now (b) says $\exists B_x \in \mathcal{B}$ such that

$$x \in B_x \subseteq B_{\alpha'} \cap B_{\beta'} \subseteq G \cap H.$$

It follows that $\bigcup_{x \in G \cap H} B_x$ contains each element of $G \cap H$ but is contained in $G \cap H$ . . . that is, $G \cap H$ *is* this union, and it belongs to τ. By induction, *finite* intersections of members of τ are always members of τ.

So τ is a topology. And it is now just about immediate from the definition of τ that $\mathcal{B}$ is a base for τ.

48. (i) Suppose that $\{B_n : n \in \mathbb{N}\}$ is a countable base for completely separable (X, τ).

Let p be any point of X.

Pick out *those* B_n's that happen to contain p, say,

$$B_{n_1}, B_{n_2}, B_{n_3}, B_{n_4}, \ldots$$

(a) They are a (countable) sequence of (open) neighbourhoods of p.

(b) Whenever N is a neighbourhood of p, $\exists$ open G such that $p \in G \subseteq N$; and 5.3 tells us that there is a member of the base . . . necessarily one of the B_{n_k}'s . . . such that $p \in B_{n_k} \subseteq G \subseteq N$.

Unfortunately, there is no guarantee so far that

$$B_{n_1} \supseteq B_{n_2} \supseteq B_{n_3} \ldots !$$

However, now define
$B'_1 = B_{n_1}$, $B'_2 = B_{n_1} \cap B_{n_2}$, $B'_3 = B_{n_1} \cap B_{n_2} \cap B_{n_3}, \ldots$ and so on. *That* certainly forces

$$B'_1 \supseteq B'_2 \supseteq B'_3 \supseteq \ldots$$

and we still have a countable sequence of (open) neighbourhoods of p, and look again at (b):

$$p \in B'_k \subseteq B_{n_k} \subseteq G \subseteq N.$$

So we have got a countable local base at p, and X is first-countable.

(ii) Consider $(\mathbb{R}, \tau_{\text{disc}})$. Any countable subset is closed and therefore certainly *not* dense, so this space fails to be separable. Therefore it cannot be completely separable (see 5.11).

On the other hand, for any $p \in \mathbb{R}$, the (mildly ridiculous) sequence

$$\{p\}, \{p\}, \{p\}, \{p\}, \ldots$$

of neighbourhoods of p is a countable local base, so the space is first-countable.

(More easily: it is metrisable, so it has got to be first-countable.)

49. Let $f : (X, \tau) \to (Y, \sigma)$ be continuous, open and onto, where X is completely separable.

Choose a countable base $\{B_n : n \in \mathbb{N}\}$ for τ.

Then $\{f(B_n) : n \in \mathbb{N}\}$ is a countable family of subsets of Y, and they all are σ-open since f is an open map. Are they a base for σ? Because, if so, (Y, σ) is completely separable and we are finished.

Let $y \in G \in \sigma$.

Choose $x \in X$ such that $f(x) = y$. Then $x \in f^{-1}(G)$, which is τ-open, so $\exists n \in \mathbb{N}$ for which

$$x \in B_n \subseteq f^{-1}(G).$$

Thus $y = f(x) \in f(B_n) \subseteq ff^{-1}(G) \subseteq G$

$$\ldots \text{ therefore } \quad y \in f(B_n) \subseteq G.$$

Hence the sets $f(B_n)$ for $n \in \mathbb{N}$ do indeed form a base. (We are using 5.3 all the time, virtually as a working definition of 'base' .)

50. Referring to the Arens–Fort space X^+, we saw that:

(a) every neighbourhood of $(0, 0)$ includes points of $\mathbb{N} \times \mathbb{N}$, so $(0, 0) \in \overline{\mathbb{N} \times \mathbb{N}}$; and yet

(b) no sequence in $\mathbb{N} \times \mathbb{N}$ can converge to $(0, 0)$.

Therefore X^+ cannot be first-countable.

In view of Essential Exercise 48(i), this tells us that X^+ cannot be completely separable.

On the other hand, $(X^+, \tau_{\text{disc}})$ is a countable set, therefore certainly separable; and because it is metrisable, it is also completely separable. (Recall that for metric spaces, separability is equivalent to complete separability) (5.11, 5.12).

Now the identity map $id_{X^+} : (X^+, \tau_{\text{disc}}) \to (X^+, \tau_{\text{af}})$, with τ_{af} being the Arens–Fort topology, is continuous (its domain is discrete, after all!) and onto, from a completely separable space to a non-completely separable space.

51. Put $\tau = \{\emptyset$ and all possible unions of finite intersections of members of $\mathcal{S}\}$. Consider an intersection of two such sets:

$$G \cap H = \bigcup_{\alpha \in A} \left(\bigcap_{j=1}^{n_\alpha} S_{\alpha,j} \right) \cap \bigcup_{\beta \in B} \left(\bigcap_{j=1}^{m_\beta} S_{\beta,j} \right).$$

If this is non-empty, *each* x in this set falls into

$$\bigcap_{j=1}^{n_{\alpha'}} S_{\alpha',j} \quad \text{and} \quad \bigcap_{j=1}^{m_{\beta'}} S_{\beta',j}$$

for some choices of α' and β'. The intersection of these two objects is a finite intersection – call it I_x – of members of $\mathcal{S}$, and it lies inside $G \cap H$. It follows that $\bigcup_{x \in G \cap H} I_x$ is exactly $G \cap H$. So $G \cap H \in \tau$, and induction tells us that τ is closed under finite intersections. It is obviously closed under unions and it includes $\emptyset$ and X, so it is a topology.

Each member S of $\mathcal{S}$ can be (perhaps rather stupidly) written as

$$(S \cap S) \cup (S \cap S)$$

so, for that reason at least, $\mathcal{S} \subseteq \tau$. Now, by construction of $\tau, \mathcal{S}$ is a subbase for τ.

If $\mathcal{S} \subseteq$ some topology τ', then each member of $\mathcal{S}$ is τ'-open and so are all finite intersections and unions therefrom (since τ' *is* a topology). So $\tau \subseteq \tau'$.

Conversely, if $\tau \subseteq \tau'$, recall that each member of the subbase $\mathcal{S}$ has to be in τ already. So $\mathcal{S} \subseteq \tau'$.

55. For each $i \in I$, the projection $\pi_i : X \to X_i$ is continuous, open and onto. But separability and compactness are preserved by continuous surjections, whereas local compactness and complete separability are preserved by continuous open surjections. Therefore π_i does all four:

$$X \text{ separable} \Rightarrow \pi_i(X) = X_i \text{ separable,}$$
$$X \text{ compact} \Rightarrow \pi_i(X) = X_i \text{ compact,}$$
$$X \text{ locally compact} \Rightarrow \pi_i(X) = X_i \text{ locally compact,}$$
$$X \text{ completely separable} \Rightarrow \pi_i(X) = X_i \text{ completely separable.}$$

56. (i) Define $\theta : X \to \Gamma(f)$ in the following (and the only reasonable) way, as shown in the figure:

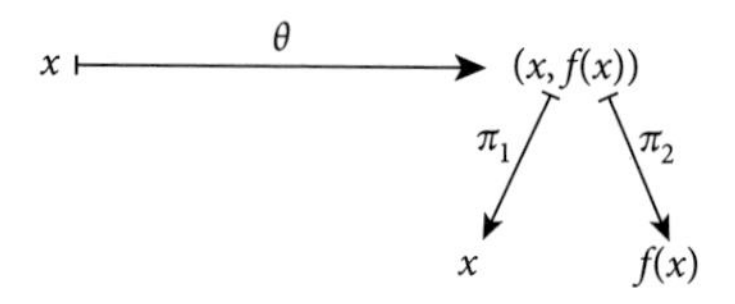

Mapping the domain of f to the graph of f.

Then

$$\pi_1 \circ \theta \text{ is } id_X, \text{ therefore continuous,}$$
$$\pi_2 \circ \theta \text{ is } f, \text{ given continuous.}$$

Therefore θ is continuous into $X \times Y$.

(Small print!) Now, by the rather forgettable Lemma 3.6, the 'co-restriction' θ *as a map onto* $\Gamma(f)$ is still continuous.

What is the inverse of θ? It has to take $(x, f(x))$ to $x \ldots$ so it is π_1! Or rather, it is the restriction of π_1 to $\Gamma(f)$: but restrictions of continuous maps are continuous, so this is continuous.

Now the maps θ and $\pi_1|_{\Gamma(f)}$ are continuous and mutually inverse, and are therefore homeomorphisms.

(ii) Now 5.23 is just the special case where f is a *constant* function.

58. (i) Let (X, τ) be T_2.
For each $(x, y) \in X^2 \setminus \Delta, x \neq y$.

Therefore $\exists$ disjoint open G, H such that $x \in G, y \in H$.

Now $G \times H$ is an open-box neighbourhood of (x, y) in X^2 and $(G \times H) \cap \Delta = \emptyset$.
That is, $X^2 \setminus \Delta$ is a neighbourhood of each of its own points.

Therefore $X^2 \setminus \Delta$ is open.

Therefore Δ is closed.

(ii) Let Δ be closed in X^2.
Given $x \neq y$ in X, $(x, y) \notin \Delta$.

Therefore $X^2 \setminus \Delta$ is a neighbourhood of (x, y)

so there is an open set, and therefore an open box $G \times H$, containing (x, y) and contained in $X^2 \setminus \Delta$.

This gives $x \in G, y \in H, G \cap H = \emptyset$,

so X is T_2.

59. Of course, if I is only a two-element set, we know this to be true: it is 6.8 (Corollary 2).

It is fairly easy to extend this by induction to show that, if I is *finite*, the conclusion is still valid; but what if I is infinite?

In $\mathbb{R}$ with τ_{usual}, put $C_n = \{1/n\}$ for $n \geq 1$ and $C_0 = \{0\}$. Each C_i is a singleton, so it is certainly compact, and the different C_i's are disjoint. But if we suppose there to exist pairwise disjoint open $\{G_i\}_{i=0,1,2,3,4,\ldots}$ such that $C_i \subseteq G_i$ for all relevant i, then

$$0 \in \text{ (open) } G_0,$$

so (since $1/n \to 0$!) many of the C_n's are actually contained in G_0, so G_n cannot always be disjoint from G_0. The assertion is therefore *false*.

60. Consider the identity map $id_X \colon (X, \tau_1) \to (X, \tau_2)$.

For all $G \in \tau_2$,

$$id_X^{-1}(G) = G \in \tau_1,$$

so id_X is continuous as well as one-to-one and onto.

By 6.9, it is a homeomorphism.

In particular, it is an open map, so

$$H \in \tau_1 \Rightarrow id_X(H) \in \tau_2 \Rightarrow H \in \tau_2,$$

that is, $\tau_1 \subseteq \tau_2$ also. Hence $\tau_1 = \tau_2$.

61. If (X, τ) is not T_2, there exist p, q in X such that $p \neq q$ and yet every open set containing p overlaps every open set containing q. Therefore each neighbourhood of p and each neighbourhood of q must contain points in common.

Choose countable local bases $\{P_n : n \in \mathbb{N}\}$ at p and $\{Q_n : n \in \mathbb{N}\}$ at q, respectively.

For each $n \in \mathbb{N}$, choose a point x_n that belongs to both P_n and Q_n. Now we have a sequence $(x_n)_{n \in \mathbb{N}}$. Does it converge to p? (1)

For each neighbourhood H of p, find a positive integer n_0 such that $P_{n_0} \subseteq H$. We see that $n \geq n_0$ implies $x_n \in P_n \cap Q_n \subseteq P_n \subseteq P_{n_0} \subseteq H$. So the answer to (1) is *yes*.

Likewise, this same sequence converges to $q\,(\neq p)$ also.

For the second part consider, for example, $(\mathbb{N}, \tau_{\text{cf}})$.

We noted in 6.5(iii) that this space is not T_2.

Yet for any $x \in \mathbb{N}$ the set of finite subsets of $\mathbb{N}$ that exclude x is countable, and may therefore be expressed as a sequence $(F_i)_{i \in \mathbb{N}}$.

Then the sequence $(\mathbb{N} \setminus F_i)_{i \in \mathbb{N}}$ is the family of all (open) neighbourhoods of x.

Using Essential Exercise 47, we produce from this a countable local base at x. Hence $(\mathbb{N}, \tau_{cf})$ is first-countable.

63. Let (X, τ_{af}) denote the Arens–Fort space. It is immediate from its definition that the complement of any single point – whether it be the exceptional point $(0, 0)$ or an 'ordinary' point – is open. Hence the space is at least T_1.

Now it is convenient to use Essential Exercise 62. Suppose we are given (in this space) a closed set K and an open set G such that $K \subseteq G$.

Does the exceptional point belong to K or not?

If $(0, 0) \notin K$, then K is open as well as closed. So, choosing $H = K$, the condition

$$K \subseteq H \subseteq \overline{H} \subseteq G$$

is trivially satisfied.

If, on the other hand, $(0, 0) \in K$, then also $(0, 0) \in G$, that is, $(0, 0) \notin X \backslash G$. It follows that $X \setminus G$ is an open set and that G is closed as well as open. Choosing $H = G$ this time, we again get

$$K \subseteq H \subseteq \overline{H} \subseteq G$$

and the two cases establish via Essential Exercise 62 that (X, τ_{af}) is T_4.

It follows that the remaining separation axioms in the hierarchy – $T_2, T_3, T_{3\frac{1}{2}}$ – are also satisfied.

64. Suppose that (A, τ_A) is a subspace of a $T_{3\frac{1}{2}}$ space (X, τ).

Given $x \notin$ closed F within (A, τ_A),

choose τ-closed F' so that $F = A \cap F'$ (2.16)

and see that $x \notin F'$.

So $\exists$ continuous $f : X \to [0, 1]$ such that $f(x) = 1$ and $f(F') = \{0\}$.

The restriction $f|_A : A \to [0, 1]$ is continuous (3.5) and $f|_A(x) = 1, f|_A(F) = \{0\}$.

(Also, (A, τ_A) is T_1: 6.2(v).)

So (A, τ_A) is $T_{3\frac{1}{2}}$.

66. (i) If (M, d) is metric and Lindelöf then, for each positive rational q, the cover of M consisting of all open balls of radius q has a countable subcover $\{B(x_{q,n}, q) : n \in \mathbb{N}\}$. The family of open sets $\{B(x_{q,n}, q) : q \in \mathbb{Q}, q > 0, n \in \mathbb{N}\}$ is countable and – very much as in the proof of 5.12 – is seen to be a base for the topology.

(ii) Let (X, τ) be compact metrisable. Choose a metric d on X such that τ_d is τ. Compact implies Lindelöf (see Essential Exercise 65), so (i) shows (X, d) to be completely separable. Completely separable implies separable (see 5.11), whence the result.

67. (i) Given $x \neq y$ in M, put $\varepsilon = \frac{1}{2}d(x, y) > 0$. Since $\{x_n : n \geq 1\}$ is dense, we can choose $n \in \mathbb{N}$ such that $d(x, x_n) < \varepsilon$. By the triangle inequality,

$$2\varepsilon = d(x, y) \leq d(x, x_n) + d(x_n, y),$$

so $d(x_n, y) > \varepsilon$. In particular, $d(x, x_n) \neq d(y, x_n)$ and $\theta(x) \neq \theta(y)$.

It is an easy consequence of the triangle inequality that, in any metric space (M, d) and for any choice of $t \in M$, the function $x \mapsto d(x, t)$ is continuous from (M, d) to the real line. In the current exercise, for any $n \in \mathbb{N}$ the composite map $\pi_n \circ \theta$ is the map $x \mapsto d(x, x_n)$ and is therefore continuous. By 5.34, θ itself is continuous.

(ii) Given compact metrisable (X, τ), choose a metric d on X such that τ_d is τ. Then (X, d) is bounded (4.19) by some positive constant K. Define $D : X^2 \to \mathbb{R}$ by $D(x, y) = d(x, y)/K$. Trivially, D is a metric on X, and it is easy to see that it induces the same topology as d did (because a set contains an open ball of *some* positive radius around a point *using the metric D* if and only if it does so *using the metric d*). Now (i) shows that there is a continuous one-to-one surjective mapping

$$\theta : (X, \tau) \to \theta(X) \subseteq [0, 1]^{\mathbb{N}}.$$

Since (X, τ) is compact and $\theta(X)$ is T_2 (via 6.5(iv) and 6.4(v)), Proposition 6.9 says that θ is a homeomorphism of X onto $\theta(X)$.

70. (i) $[0, 1]$ is T_2 (because metric) and T_2-ness is productive (6.5 (v)), so any product of (copies of) $[0, 1]$ is T_2.

(ii) Tychonoff's theorem tells us that $\prod Y_\alpha$ is compact, since each $Y_\alpha = [0, 1]$ is compact.

(iii) For each $\alpha \in I$, consider the αth projection

$$\pi_\alpha : \prod Y_\alpha \to Y_\alpha = [0, 1].$$

We see that

$$\begin{aligned}(\pi_\alpha \circ \theta)(x) &= \text{ the } \alpha^{\text{th}} \text{ coordinate of } \theta(x)\\ &= f_\alpha(x) \quad \forall x \in X,\end{aligned}$$

that is, $\pi_\alpha \circ \theta = f_\alpha$, which is continuous.

Now 5.34 says that θ is continuous.

(iv) For $x \neq y$ in $X, x \notin$ closed $\{y\}$, so $T_{3\frac{1}{2}}$ says $\exists$ continuous $f : X \to [0, 1]$ such that $f(x) = 0$ and $f(y) = 1$. Now, f is one of the f_α's and the αth coordinates of $\theta(x)$ and $\theta(y)$ are 0 and 1, respectively, so certainly $\theta(x) \neq \theta(y)$. Hence θ is one-to-one.

72. For each $x \in W \setminus \{t\}$, there is a least element (x^+, say) that is greater than x. Then $[0, x] = [0, x^+)$. This shows firstly that $[0, x^+)$ is open in the intrinsic topology and includes x . . . so it is actually a neighbourhood of x . . . and secondly that it is compact: because it is a well-ordered set with a maximum element x . . . see 6.28.

(Actually, there is a subtlety here, since 6.28 says that $[0, x]$ under *its own* intrinsic topology is compact, whereas we want $[0, x]$ with the *subspace* topology from $W \setminus \{t\}$'s intrinsic topology to be compact. The conscientious reader will want to check that these topologies are actually equal.)

74. (a) A rather lazy example of a sequence in $(W \times Y) \setminus \{(t, 1)\}$ that converges to $(t, 1)$ is

$$(t, 0), \left(t, \frac{1}{2}\right), \left(t, \frac{2}{3}\right), \left(t, \frac{3}{4}\right), \left(t, \frac{4}{5}\right), \ldots$$

(We are taking it as read that, in a product space $X \times Y$, a sequence $\left((x_n, y_n)\right)_{n\geq 1}$ converges to $(l, m) \Leftrightarrow x_n \to l$ and $y_n \to m$. You can easily prove this as an additional exercise.)

(b) Suppose that there is a sequence $\left((x_n, y_n)\right)_{n\geq 1}$ in $(W \times Y) \setminus (\{t\} \times Y)$ converging to $(t, 1)$. Notice in particular that every x_n must belong to $W \setminus \{t\}$. But π_1 is continuous (and preserves limits of sequences), so $x_n \to t$. This contradicts 6.29(i).

75. For any $(x, -x)$ on the line $L = \{(x, -x) : x \in \mathbb{R}\}$ inside SP,

$$N = [x, x + 1) \times [-x, -x + 1)$$

is an open box . . . therefore an open set in SP . . . containing $(x, -x)$ but no other point on L. That is,

$$N \cap L = \{(x, -x)\}$$

is open in the subspace topology that SP gives L.

Therefore L is discrete!

Since L is uncountable and discrete, it is *not* completely separable.

But (5.10) complete separability is hereditary, so SP cannot be completely separable either.

Yet every open box $[a, b) \times [c, d)$ in SP contains the 'Euclidean' open box $(a, b) \times (c, d)$, which certainly includes some point whose coordinates are both rational. It follows that $\mathbb{Q} \times \mathbb{Q}$ is dense in SP just as it is dense in $\mathbb{R}^2$ under its usual topology. Since $\mathbb{Q} \times \mathbb{Q}$ is countable, SP is separable.

77. (i) (One solution is to modify the proof of 3.11.) Take the map θ from $Cantor_3$ to $[0, 1]$ defined by

$$\theta\left(\sum \frac{a_n}{3^n}\right) = \sum\left(\frac{a_n/2}{2^n}\right)$$

(remembering that each a_n is either 0 or 2).

Given x in $Cantor_3$ and $\varepsilon > 0$, take a value of n such that 2^{-n} is less than ε;

then $y \in Cantor_3$ and $|x - y| < 3^{-n}$ together guarantee that y belongs to the same stage-n block in the construction of $Cantor_3$ as x does;

therefore $\theta(y)$ and $\theta(x)$ possess the same first n binary digits, whence $|x - y| \leq 2^{-n} < \varepsilon$ and continuity of θ is established. The fact that each number in $[0, 1]$ has a binary expansion shows θ to be onto.

(ii) Using (i), take a continuous surjection $g : Cantor_3 \to [0, 1]$. Define $G : (Cantor_3)^{\mathbb{N}} \to ([0, 1])^{\mathbb{N}}$ via $G((x_n)_{n\in\mathbb{N}}) = (g(x_n))_{n\in\mathbb{N}}$. Then G is surjective because g is, and G is continuous 'for the usual reasons' involving 5.34 (if π_n and π'_n denote the nth projections in its range and its domain, respectively, then $\pi_n \circ G = g \circ \pi'_n$ and is therefore continuous for all n).

(iii) In view of 6.10(ii), there is a homeomorphism $\theta : Cantor_3 \to (Cantor_3)^{\mathbb{N}}$. Then, with G as in (ii), $G \circ \theta$ is a continuous surjection of $Cantor_3$ onto $[0, 1]^{\mathbb{N}}$.

78. (i) For each x in K, $k(x)$ is certainly the same point as x, so k is a map *onto* K. Furthermore, in this case, given $\varepsilon > 0$ and $y \in C \cap (x - \varepsilon/2, x + \varepsilon/2)$, $|y - k(y)|$ must be $\leq |y - x| < \varepsilon/2$, so $|k(y) - k(x)| \leq |k(y) - y| + |y - k(x)| < \varepsilon/2 + \varepsilon/2 = \varepsilon$. Hence k is continuous at x.

Now consider the case where $x \notin K$ and $x < k(x)$. Put $\varepsilon = k(x) - x$ and, bearing in mind that $k(x)$ is *uniquely* closest to x, find $\delta > \varepsilon$ such that the interval $(x - \delta, x)$ contains no element of K. For any $y \in C$ between $x \pm (\delta - \varepsilon)/2$, the closest point in K to y is still $k(x)$. Therefore k is constant on a C-neighbourhood of x and, again, continuous there. (The remaining case – that in which $x \notin K$ and $x > k(x)$ – is similar.)

(ii) According to 1.11(i), the point midway between two distinct elements of $Cantor_5$ never lies in $Cantor_5$, so any closed subset K of this space satisfies the uniqueness condition in (i) above. Now the claimed result is immediate from (i).

79. (Since all the Cantor spaces are homeomorphic (3.11), we are free to choose any arithmetical base that suits us.)

By Essential Exercise 67, there is no loss of generality in assuming that X is actually a subset of $[0, 1]^{\mathbb{N}}$.

From Essential Exercise 77, there is a continuous surjective $G : Cantor_5 \to [0, 1]^{\mathbb{N}}$.

Then the set $K = G^{-1}(X)$ is closed in $Cantor_5$, so Essential Exercise 78 tells us $\exists \phi : Cantor_5 \to K$ that is continuous and onto.

Now $G \circ \phi$ continuously maps the Cantor space onto X as required.

Suggestions for further reading

Dixmier, J. *General Topology*. Springer (2010).
Gemignani, M. *Elementary Topology* (2nd edn). Dover Publications (1990).
Kasriel, R. *Undergraduate Topology*. Dover Publications (2009).
Lipschutz, S. *Schaum's Outline of General Topology*. McGraw-Hill (1988).
Mendelson, B. *Introduction to Topology*. Dover Publications (1990).
Munkres, J. R. *Topology*. Prentice Hall (2000).
Steen, L. A. and Seebach, J. A. *Counterexamples in Topology*. Dover Publications (1995).
Sutherland, W. A. *Introduction to Metric and Topological Spaces* (2nd edn). Oxford University Press (2009).

Index